SpringerBriefs in Molecular Science

Chemistry of Foods

W0080044

Series Editors

Salvatore Parisi, Lourdes Matha Institute of Hotel Management and Catering Technology, Thiruvananthapuram, Kerala, India

Ricardo Pereira, Centre of Biological Engineering, University of Minho, Braga, Portugal

The series Springer Briefs in Molecular Science: Chemistry of Foods presents compact topical volumes in the area of food chemistry. The series has a clear focus on the chemistry and chemical aspects of foods, topics such as the physics or biology of foods are not part of its scope. The Briefs volumes in the series aim at presenting chemical background information or an introduction and clear-cut overview on the chemistry related to specific topics in this area. Typical topics thus include:

- Compound classes in foods—their chemistry and properties with respect to the foods (e.g. sugars, proteins, fats, minerals, …)
- Contaminants and additives in foods—their chemistry and chemical transformations
- Chemical analysis and monitoring of foods
- Chemical transformations in foods, evolution and alterations of chemicals in foods, interactions between food and its packaging materials, chemical aspects of the food production processes
- Chemistry and the food industry—from safety protocols to modern food production

The treated subjects will particularly appeal to professionals and researchers concerned with food chemistry. Many volume topics address professionals and current problems in the food industry, but will also be interesting for readers generally concerned with the chemistry of foods. With the unique format and character of SpringerBriefs (50 to 125 pages), the volumes are compact and easily digestible. Briefs allow authors to present their ideas and readers to absorb them with minimal time investment. Briefs will be published as part of Springer's eBook collection, with millions of users worldwide. In addition, Briefs will be available for individual print and electronic purchase. Briefs are characterized by fast, global electronic dissemination, standard publishing contracts, easy-to-use manuscript preparation and formatting guidelines, and expedited production schedules.

Both solicited and unsolicited manuscripts focusing on food chemistry are considered for publication in this series. Submitted manuscripts will be reviewed and decided by the series editor, Prof. Dr. Salvatore Parisi.

To submit a proposal or request further information, please contact Dr. Sofia Costa, Publishing Editor, via sofia.costa@springer.com or Prof. Dr. Salvatore Parisi, Book Series Editor, via drparisi@inwind.it or drsalparisi5@gmail.com

Vanessa Alves · Guilherme de Figueiredo Furtado ·
Gabriela Alves Macedo

Chemical and Enzymatic Interesterification for Food Lipid Production

 Springer

Vanessa Alves
Food Science and Nutrition Department
State University of Campinas
Campinas, São Paulo, Brazil

Guilherme de Figueiredo Furtado
Centre of Natural Sciences
Federal University of São Carlos
Buri, Brazil

Gabriela Alves Macedo
Food Science and Nutrition Department
State University of Campinas
Campinas, São Paulo, Brazil

ISSN 2191-5407   ISSN 2191-5415 (electronic)
SpringerBriefs in Molecular Science
ISSN 2199-689X   ISSN 2199-7209 (electronic)
Chemistry of Foods
ISBN 978-3-031-67404-4   ISBN 978-3-031-67405-1 (eBook)
https://doi.org/10.1007/978-3-031-67405-1

This Springer imprint is published by the registered company Springer Nature Switzerland AG
The registered company address is: Gewerbestrasse 11, 6330 Cham, Switzerland

If disposing of this product, please recycle the paper.

Contents

Chapter 1
Chemical and Enzymatic Interesterification for Food Lipid Production: An Introduction

1.1 Introduction

Lipids are naturally present in nature and, consequently, in foods, in the form of emulsions and/or oils and fats. They are a fundamental part of the diet, playing essential physiological and metabolic roles in the human body. Additionally, lipids contribute to the organoleptic perceptions and physicochemical properties of foods [1, 2]. Structurally, lipids have between 4 and 24 carbon atoms, primarily consisting of fatty acids (FAs) with an aliphatic chain and a carboxylic group [3]. Saturated FAs, monounsaturated (MUFAs), and polyunsaturated (PUFAs) are structurally arranged at positions *sn*-1, *sn*-2, and *sn*-3 of the glycerol molecule [4]. Their positioning in the triacylglycerol (TAG) molecule can determine the functional and physical properties of the lipid, as well as its metabolic functions and nutritional benefits [5]. The proportions of these FAs may vary depending on the type of lipid matrix provided [6]. In vegetable oils, for example, there is a predominance of saturated FAs at positions *sn*-1 and *sn*-3, while unsaturated FAs are prevalent at position *sn*-2 [7].

In this context, natural lipids necessitate modifications since their desirable properties and applications depend not only on their composition but also on the structure of TAGs. Therefore, the partial hydrogenation of vegetable oils has been employed as one of the primary methods on a large scale for lipid modification, gaining significant importance in the food industry, especially in the 1950s [8]. This process has been applied for various technological purposes in a wide range of foods, providing unique texture and stability. It transforms vegetable oils rich in unsaturated FAs into plastic fats (semisolids), thereby enhancing the textural and oxidative stability characteristics of products such as margarines, spreads, cookie fillings, ice creams, among others [9]. Partial hydrogenation involves the addition of hydrogen atoms to cis-unsaturated FAs, resulting in the formation of trans-isomerized FAs [10].

From the 1960s onward, there has also been a growing concern about replacing saturated fats of animal origin, with recommendations to limit their consumption

V. Alves et al., *Chemical and Enzymatic Interesterification for Food Lipid Production*, Chemistry of Foods, https://doi.org/10.1007/978-3-031-67405-1_1

[11]. Despite lipids being one of the main dietary macronutrients, their excessive intake, particularly of saturated FAs, is associated with a risk factor for obesity and the exacerbation of metabolic syndromes [12, 13]. The unrestricted and excessive consumption of trans-FAs in the Western diet in recent years has led to undesirable health effects overall. Advances in research have raised concerns about the negative impact of trans-FAs on health. Their consumption has been linked to detrimental health effects similar to those caused by the excessive consumption of saturated FAs from animal sources, contributing to the development of cardiovascular diseases (CVD), influencing in includes the insulin-mediated storage of lipids in muscles, an increase in low-density lipoprotein (LDL) cholesterol levels, and a decrease in high-density lipoprotein (HDL) cholesterol levels. These factors have also been implicated in the incidence of hyperlipidemia, diabetes, cancer, and other non-communicable chronic diseases (NCDs) [9, 14].

Based on these concerns, legislations and guidelines from health and food-related organizations emerged, aiming to restrict the consumption of partially hydrogenated fats [8, 15, 16]. In 2006, the Food and Drug Administration (FDA) mandated food industries to disclose the amount of trans-FAs in their labels [15]. In 2015, the FDA removed partially hydrogenated fats from the Generally Recognized as Safe (GRAS) category and prohibited their addition to foods after June 2018 [17]. In 2020, the Pan American Health Organization (PAHO), in conjunction with the World Health Organization (WHO), published an action plan to eliminate industrially produced trans-FAs [16]. In Brazil, the National Health Surveillance Agency (ANVISA) issued RDC 332/2019 in 2019, banning the use of partially hydrogenated fats from January 2023 [18].

With the proposition to restrict and/or limit the use of saturated fats and the understanding of the negative impact of trans-FAs on health, coupled with the prohibition of their use, the food industry had to adopt alternative technologies. Lipid interesterification or lipid synthesis, also known as structured lipids (SLs), emerged as a viable and promising method for the production of dietary lipids without trans-isomer content [19, 20]. SLs are modified or restructured TAGs obtained through chemical and/or enzymatic interesterification or transesterification, or even through genetic engineering methods [21]. During the interesterification synthesis, catalyzed by specific agents, ester bonds undergo random cleavage, resulting in free FAs that are distributed, reorganized, and rearranged in different positions in the glycerol molecule without altering its lipid composition [6, 22].

Thus, restructuring lipids by altering the profiles of FAs, including levels of unsaturation, chain length, and positional distribution, can lead to improved physicochemical characteristics of plastic fats. These improvements include crystallization, melting point, polymorphism, solid fat content, viscosity, and consistency, as well as oxidative stability when compared to unmodified lipids [23]. Additionally, such modified lipids may offer satisfactory and desirable nutritional functions, promoting a faster energy source that could be beneficial for individuals with disorders in lipid absorption and metabolism, and supporting anti-obesity actions [6, 24]. Moreover, the consumption of SLs may contribute to public health by increasing the intake of

MUFAs and PUFAs, as well as FAs with functional properties, such as docosa-hexaenoic acid (DHA), eicosapentaenoic acid (EPA), and behenic acid [4, 25]. This substitution can efficiently and cost-effectively replace partially hydrogenated vegetable oils.

1.2 Production of Structured Lipids

1.2.1 Chemical Interesterification

Chemical interesterification is characterized as a completely random reaction, resulting in a random redistribution of acylglycerols to produce a predetermined composition of TAGs until reaching thermodynamic equilibrium among all possible combinations, however, the reaction lacks selectivity, making it impossible to control the distribution of FAs in the glycerol molecule's positions (Fig. 1.1) [26].

This synthesis is commonly employed by the food industry as it is feasible for the lipid modification of oils and fats, offering lower investment costs, applicability, and scalability for the large-scale production of SLs [27]. This is because these reactions are typically catalyzed by alkali metal compounds and alkali metal alkoxides, which are relatively inexpensive and readily available, these catalysts require high reaction temperatures (90–150 °C), enabling rapid reactions in approximately 20–30 min [6]. Presumably, during chemical interesterification, the catalytic material consists of anions formed in the lipid substrate after the addition of an alkaline precursor [28]. Commonly employed chemical catalysts include sodium alkoxides, metallic sodium, and occasionally mixtures of sodium/potassium, as well as sodium or potassium hydroxides in combination with glycerol [4, 29]. However, alkali hydroxides, such as sodium hydroxide, are less active and therefore require higher reaction temperatures [28].

It is noteworthy that chemical interesterification has some disadvantages: random-ization is one issue, and the reaction must be halted for catalyst inactivation by adding

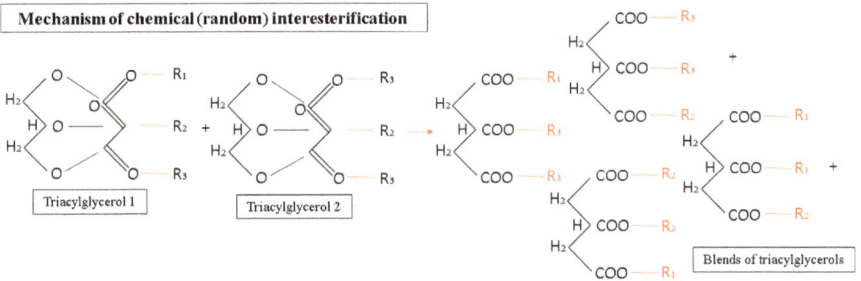

Fig. 1.1 Diagram of the random mechanism of chemical interesterification. Figure adapted from Ref. [4]

water or organic acid, converting methyl esters into free FAs, this process results in wasted catalyst that cannot be reused, leading to the generation of polluting effluents, additionally, the catalyst used in chemical interesterification is highly reactive and toxic, significant amounts of reaction intermediates can be obtained, such as FA esters, soaps, free FAs, mono- and diacylglycerides (MAGs and DAGs), tocopherols, and esterified sterols, this, post-processing steps, such as bleaching and deodorization, are required to produce a product safe enough for consumption, as the final products may be contaminated by these intermediates, subsequently reducing the yield of SLs [30, 31].

The substitution of chemically catalyzed processes with enzyme-catalyzed processes is becoming more desirable as they are recognized as natural, exhibit selectivity of biocatalysts under mild reaction conditions, resulting in highly pure products with higher yields; additionally, these processes are environmentally friendly. Moreover, enzyme-catalyzed processes are considered safe for handling compared to chemical catalysts [27, 31]. In this context, there are currently few reports of studies focusing solely on chemical interesterification, with the more common approach being the comparison between chemical and enzymatic interesterification processes. Table 1.1 presents some examples available in the literature.

1.2.2   Chemical Reaction Mechanism

The mechanism of the chemical interesterification reaction occurs in two steps, with the initial formation of a FA catalyst followed by the propagation of the reaction. Some studies suggest that the involved anion is a glycerolate [28]; however, there are reports of shortcomings in this mechanism, leading to the proposal of an enolate mechanism [41]. Both mechanisms are discussed below.

In the glycerolate mechanism in the reaction catalyzed by sodium methoxide ($NaOCH_3$), the nucleophilic methoxide ion (CH_3O^-) acts by attacking the carbonyl atom ($^+$) in the TAG to form a tetrahedral intermediate (diacylglycerol anion DAG) [42]. Subsequently, the DAG acts by attacking another active ester linkage in the TAG, forming an intermediate complex, then, the DAG and the newly formed TAG molecule dissociate before completing the interesterification process, in a recurring process until all positions of the acyl groups are altered randomly. In other words, this process occurs continuously until thermodynamic equilibrium is reached, making it impossible to control the reaction and obtain TAGs with a target structure (position specificity), even when controlling the synthesis process conditions [43, 44].

The enolate mechanism occurs when the enolate anion reacts with (free) hydroxyl groups in partial glycerides, leading to the interesterification reaction [28]. When the catalyst sodium methoxide ($NaOCH_3$) is used in the reaction, it can react in various ways, giving rise to a series of different anions, such as enolate ($CH_2COCH_2^-$), for example, which contains a pair of electrons in the carbon–oxygen double bond, conferring nucleophilic properties. At the moment when sodium methoxide reacts with the residual water resulting from the catalyst deactivation process, it releases

Table 1.1 Chemical interesterification for the production of structured lipids

Reaction type	Sources of lipid	Catalyst	Conditions	Objectives	Reference
Chemical and enzymatic interesterification	Hydrogenated canola oil + high oleic sunflower oil	Sodium methoxide + *Candida antarctica* lipase	0.3% at 88 °C + 5% enzyme at 68 °C	Compared the physicochemical properties, microstructure, polymorphism, and rheological properties among the different syntheses	[32]
Chemical interesterification	Soybean oil and fully hydrogenated soybean oil	Sodium methoxide	0.4% at 100 °C	Evaluated thermal behavior, microstructure, polymorphism, and crystallization properties	[33]
Chemical interesterification	Palm stearin + olive oil	Sodium methoxide	0.5% at 70 °C	Investigated the improvement in the physical properties of SLs with palm stearin and olive oil	[34]
Chemical and enzymatic interesterification	Rice bran oil, shea olein, and palm stearin + rosemary extract	Sodium methoxide + Lipozyme TL IM	0.2% at 105 °C + 10% enzyme at 70 °C	Compared the physicochemical properties and oxidative stability in different syntheses for application in margarines	[35]
Chemical interesterification	Soybean oil + crambe hardfat	Sodium methoxide	0.4% at 100 °C	Evaluated the physicochemical characteristics and texture of SLs with low trans-fat content	[36]
Chemical and enzymatic interesterification	Fully hydrogenated sunflower oil + high oleic sunflower oil	Sodium methoxide + Lipozyme TL IM	0.4% at 100 °C + 7% enzyme at 70 °C	Compared the regiospecific distribution of TAGs after chemical and enzymatic synthesis using C^{13} nuclear magnetic resonance	[37]
Chemical and enzymatic interesterification	*Acer truncatum* oil, palm stearin, and palm kernel oil	Sodium methoxide + Lipozyme RM IM	0.4% at 105 °C + 8% enzyme at 60 °C	Produced trans-free SLs for margarines, rich in nervonic acid, a FA with beneficial effects on mental health	[38]
Chemical and enzymatic interesterification	Refined palm oil + palm kernel oil	Sodium methoxide + Lipozyme TL IM	0.2% at 110 °C + 4% enzyme at 70 °C	Compared crystallization behavior, microstructure, polymorphism, and thermal properties among different syntheses	[27]

(continued)

Table 1.1 (continued)

Reaction type	Sources of lipid	Catalyst	Conditions	Objectives	Reference
Chemical and enzymatic interesterification	Palm olein + palm kernel oil + palm stearin	Sodium methoxide + Lipozyme TL IM	0.3% at 105 °C + 813.0 g enzyme at 60 °C	Compared the physicochemical properties and crystallization among different interesterification processes	[39]
Chemical and enzymatic interesterification	Olive oil + soybean oil + hydrogenated crambe oil	Sodium methoxide + Lipozyme TL IM	0.4% at 100 °C + 10% enzyme at 60 °C	Compared both syntheses to obtain low-calorie SLs	[40]

free methanol in the same proportion as the catalyst, and this methanol will react with the enolate to form FA methyl esters. Additionally, the residual water will react with the enolate to form free FAs, also in the same proportion as the catalyst, and a glycerolate anion. This glycerolate anion will abstract an α-hydrogen from a portion of FA and restructure the enolate anion [45]. This mechanism explains why the reaction mixtures of chemical interesterification contain proportions of methyl esters of FAs and free FAs equivalent to the amount of sodium methoxide added, to remove these, washings with water or hydrogel silica adsorption are carried out, and additional processes such as deodorization and bleaching are still necessary [31]. The enolate mechanism also explains why reactions subjected to temperatures above 100 °C exhibit lower catalytic activity, due to the formation of a β-ketoester from the enolate anion [45]. For the chemical reaction to occur efficiently, it is recommended to consider the percentage necessary for catalysis plus the amount of sodium methoxide needed to neutralize the free FAs and the methyl esters FAs plus peroxides present in the lipid substrate, thus, [46] recommends 0.1% sodium methoxide to neutralize 0.05% free FAs and 0.054% sodium methoxide to neutralize 0.05% free FAs and 0.054% sodium methoxide to neutralize 1.0% peroxides.

1.2.3 Enzymatic Interesterification

SLs can be synthesized through enzymatic reactions involving acidolysis, transesterification, or alcoholysis, and/or interesterification [47] (Fig. 1.2). Acidolysis is a reaction between an ester and a free FA, requiring an acid–base catalyst to form MAGs and DAGs through the partial hydrolysis of FA esters in TAG molecules. This process results in the formation of new TAGs when the hydroxyl group of MAG and/ or DAG reacts with a free FA [6]. Acidolysis and interesterification reactions differ in the formation of the new ester: if the acyl-enzyme group comes from a free FA, it is termed acidolysis, and if derived from an ester (another glyceride), it is considered interesterification [48].

In transesterification, the ester linkages connecting FAs to glycerol are initially cleaved, forming free FAs that are rearranged randomly in a pool of FAs, these are then reesterified, either at a new position on the same glycerol molecule (intraesterification) or on another glycerol molecule (interesterification), this continuous process proceeds until a lower energy state of the enzyme is achieved, reaching a thermodynamic equilibrium, the final products of this process are new TAGs and reaction intermediates [4, 6].

Transesterification by alcoholysis occurs through the migration of the acyl group between an acylglycerol and an alcohol, resulting in the formation of a new ester and alcohol [49]. In this reaction, Brønsted acids act as catalysts, with carboxylic acids, such as sulfuric acid (H_2SO_4) or hydrochloric acid (HCl), being most commonly used. These acids provide catalytic activity to the reaction by protonating the oxygen group of the ester, forming a protonated carbonyl that is more reactive and susceptible

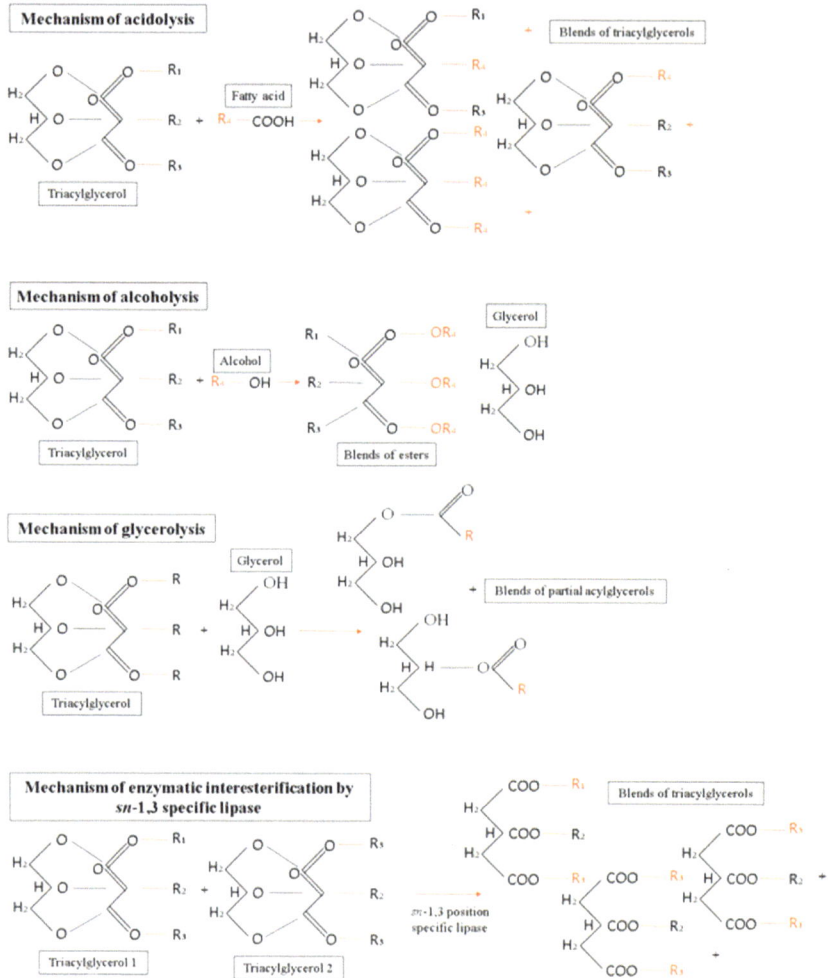

Fig. 1.2 Diagram of enzymatic reaction mechanisms for acidolysis, alcoholysis, glycerolysis, and enzymatic interesterification by specific *sn*-1,3 lipase. Figure adapted from Ref. [4]

to nucleophilic attack by the alcohol. This leads to the breaking of the carbon–oxygen bond and the formation of the new ester, followed by rearrangement in the molecule [50]. It is important to note that the Brønsted acid is not consumed in the reaction and can catalyze various stages of transesterification. This is possible because the water formed as a by-product in the synthesis deprotonates the newly formed carboxylic acid, allowing for the efficient production of esters from esters and alcohols [50]. When methanol, ethanol, or glycerol are the alcohols used, this reaction is called methanolysis, ethanolysis, and glycerolysis, respectively [47]. Regarding the production of SLs for food, it is emphasized that alcoholysis and glycerolysis reactions are associated with the excessive production of undesirable reaction by-products such as MAGs, DAGs and free FAs. On the other hand, methanolysis is not recommended due to the toxicity of methanol [4, 51, 52]. This synthesis is widely employed in biodiesel production [53].

Enzymatic interesterification catalyzed by lipases shows higher values for production compared to chemical methods; however, it has more advantages as it is capable of incorporating FAs (acyl groups) at specific positions in the glycerol structure, resulting in lipids with desirable characteristics [54]. Lipases exhibit specificity primarily at positions sn-1,3, allowing strictly directed exchange at the first and third positions of the TAG molecule, resulting in lipids with desirable characteristics that cannot be achieved through conventional chemical methods. These lipids possess improved physicochemical properties of plastic fats, including enhanced crystallization, melting point, polymorphism, solid fat content, viscosity, and consistency, additionally, they demonstrate improved oxidative stability and satisfactory nutritional functions, such as promoting a faster energy source for health benefits in individuals with disorders in lipid absorption and metabolism [27]. Moreover, they contribute to low-calorie lipids through anti-obesogenic actions [22].

Enzymatic interesterification demonstrates stability at mild temperatures ($\leq$70 °C), preserving thermolabile compounds. It is ecologically sustainable as it promotes enzymatic reuse for multiple reactions, leading to increased process efficiency through continuous operation is considered more natural and contributes to the reduction of greenhouse gas emissions, water and energy consumption, as well as the reuse of industrial waste, potentially lowering operational costs on a large scale [29, 55, 56]. However, its drawbacks include a longer reaction time and higher enzyme purchase costs compared to chemical interesterification.

1.2.4 Enzymatic Reaction Mechanism

Enzymatic interesterification involves two reaction stages, starting with acylation: In this stage, the amino acid serine, located at the enzyme's active site, initiates a nucleophilic attack on the carbonyl carbon of the substrate, forming a covalent acyl-enzyme complex [6]. On the other hand, deacylation forms a tetrahedral structure (DAG anion) which is stabilized by the formation of two hydrogen bonds [57]. During this reaction, alcohol is released through the breaking of the carbon–oxygen bond within the ester and from three amino acid residues, catalytically formed, to covalently bind with acylglycerol. Thus, interesterification is based on acylation and deacylation in a "back-and-forth" mechanism [58, 59]. In other words, through the nucleophilic attack of serine on TAG, the acyl-enzyme (without the presence of water) is obtained, and DAG acts as the nucleophile, generating a new TAG and enzyme, and this continuous mechanism persists until thermodynamic equilibrium is once again achieved, resulting in new TAGs and reaction intermediates as the final products [60].

In vegetable oils, saturated FAs are naturally distributed at positions sn-1,3 of the TAG [43]. Therefore, during the interesterification synthesis, acyl migration occurs, resulting in the randomization FAs. This process is directly related to the type of catalyst, which predominantly acts on TAGs at sn-1,3 to produce 1,2- and 2,3-DAGs, leaving the sn-2 position with minimal acyl exchanges [37]. In this way, the FA present at sn-2 of 1,2- and 2,3-DAGs can spontaneously migrate, producing more stable 1,3-DAG, which will subsequently form new TAGs with randomly rearranged FAs at positions sn-1,3, and 2 until reaching an equilibrium among all possible combinations. When lipases and specific reaction conditions are employed, the sn-2,3 (1,2)-DAG anions can attack the intermolecular carbonyl group at sn-2, producing the sn-1,3-DAG anion for carbonyl addition, altering the acyl group position at sn-2 [42]. This may be favorable for the formation of TAGs with unsaturated FAs at sn-1,3 and saturated FAs at sn-2, resulting in specifics SLs, can be applied as substitutes for human milk fat (HMF), for example [61, 62].

As already discussed in this chapter, enzymatic interesterification presents more advantageous reaction characteristics compared to chemical interesterification. Thus, this comparison of the main characteristics of each interesterification reaction is presented in Table 1.2.

Table 1.2 Comparison of the main characteristics of chemical and enzymatic interesterification

Main characteristics	Interesterification reaction	
	Chemical	Enzymatic
Catalyst types	Chemical catalysts include sodium alkoxides, metallic sodium, and occasionally mixtures of sodium/potassium, as well as sodium or potassium hydroxides in combination with glycerol	Biocatalyst for lipases (heterogeneous class of enzymes obtained from fungi, yeasts, and bacteria of animal and/or plant origin)
Syntheses	Random reaction that produces positional randomization of the acyl groups in the TAG	Regioselective, stereo-selective, selective/specific and non-specific manner (random) producing products similar to chemical interesterification
Reaction condition	Are relatively inexpensive and readily available, these catalysts require high reaction temperatures, enabling rapid reactions	Demonstrates stability at mild temperatures, preserving thermolabile compounds. Its drawbacks include a longer reaction time and higher enzyme purchase costs
Production of by-products and wastes	Intermediates can be obtained, such as FA esters, soaps, free FAs, mono- and diacylglycerides (MAGs and DAGs)	Undesirable and inevitable reaction processes can still occur, resulting in the formation of by-products
Control of reaction	The reaction must be halted for catalyst inactivation by adding water or organic acid	Promotes enzymatic reuse for multiple reactions, leading to increased process efficiency through continuous operation is considered a more natural
Technical challenge	Post-processing steps, such as bleaching and deodorization	The activity and stability of enzymes can be affected by temperature, pH, and the presence of heavy metals
Trans-fat production	Trans-free fat	Trans-free fat
Handling conditions	Catalyst is highly reactive and toxic	Enzyme-catalyzed processes are considered safe for handling
Environmental concern	Environmentally not friendly	Environmentally friendly

References [4, 29, 44, 47, 63]

References

1. Jala RCR, Ganesh Kumar C (2018) Designer and functional food lipids in dietary regimes: current trends and future prospects. In: Alternative and replacement foods, Elsevier, pp 283–316. https://doi.org/10.1016/B978-0-12-811446-9.00010-1
2. Stahl MA, Buscato MHM, Grimaldi R, Cardoso LP, Ribeiro APB (2018) Structuration of lipid bases with fully hydrogenated crambe oil and sorbitan monostearate for obtaining zero-trans/low sat fats. Food Res Int 107:61–72. https://doi.org/10.1016/j.foodres.2018.02.012
3. Bornscheuer UT, Adamczak M, Soumanou MM (2012) Lipase-catalysed synthesis of modified lipids. In: Lipids for functional foods and nutraceuticals, Elsevier, pp 149–182. https://doi.org/10.1533/9780857097965.149
4. Sivakanthan S, Madhujith T (2020) Current trends in applications of enzymatic interesterification of fats and oils: a review. LWT 132:109880. https://doi.org/10.1016/j.lwt.2020.109880
5. Mensink RP, Sanders TA, Baer DJ, Hayes K, Howles PN, Marangoni A (2016) The increasing use of interesterified lipids in the food supply and their effects on health parameters. Adv Nutr 7(4):719–729. https://doi.org/10.3945/an.115.009662
6. Kadhum AAH, Shamma MN (2017) Edible lipids modification processes: a review. Crit Rev Food Sci Nutr 57(1):48–58. https://doi.org/10.1080/10408398.2013.848834
7. Buchgraber M, Ulberth F, Emons H, Anklam E (2004) Triacylglycerol profiling by using chromatographic techniques. Eur J Lipid Sci Technol 106(9):621–648. https://doi.org/10.1002/ejlt.200400986
8. David ML, Guivant JS (2012) A gordura trans: entre as controvérsias científicas e as estratégias da indústria alimentar. Política & Sociedade 11(20). https://doi.org/10.5007/2175-7984.2012v11n20p49
9. Thirumdas R (2023) Partial hydrogenation of oils using cold plasma technology and its effect on lipid oxidation. J Food Sci Technol 60(6):1674–1680. https://doi.org/10.1007/s13197-022-05434-z
10. Gengatharan A, Mohamad NV, Zahari CNMC, Vijayakumar R (2023) Oleogels: innovative formulations as fat substitutes and bioactive delivery systems in food and beyond. Food Struct 38:100356. https://doi.org/10.1016/j.foostr.2023.100356
11. Berry SE et al (2019) Interesterified fats: what are they and why are they used? a briefing report from the roundtable on interesterified fats in foods. Nutr Bull 44(4):363–380. https://doi.org/10.1111/nbu.12397
12. Albracht-Schulte K et al (2018) Omega-3 fatty acids in obesity and metabolic syndrome: a mechanistic update. J Nutr Biochem 58:1–16. https://doi.org/10.1016/j.jnutbio.2018.02.012
13. Çakmur H (2020) Introductory chapter: unbearable burden of the diseases—obesity. In: Obesity, IntechOpen. https://doi.org/10.5772/intechopen.85234
14. Basak S, Banerjee A, Pathak S, Duttaroy AK (2022) Dietary fats and the gut microbiota: their impacts on lipid-induced metabolic syndrome. J Funct Foods 91:105026. https://doi.org/10.1016/j.jff.2022.105026
15. Food and Drug Administration. Final determination regarding partially hydrogenated oils (removing trans fat). https://www.fda.gov/food/food-additives-petitions/final-determination-regarding-partially-hydrogenated-oils-removing-trans-fat.
16. da Sopas OPA Plano de Ação para Eliminar Os Ácidos Graxos Trans De Produção Industrial 2020—2025. https://www.paho.org/en/documents/plan-action-elimination-industrially-produced-trans-fatty-acids-2020-2025
17. O'Sullivan CM, Barbut S, Marangoni AG (2016) Edible oleogels for the oral delivery of lipid soluble molecules: composition and structural design considerations. Trends Food Sci Technol 57:59–73. https://doi.org/10.1016/j.tifs.2016.08.018
18. de Brasil Anvisa AN Norma sobre gordura trans em alimentos. RDC 332/2019. https://www.gov.br/anvisa/pt-br/assuntos/noticias-anvisa/2020/publicada-norma-sobre-gordura-trans-em-alimentos

19. Xie W, Yang X, Zang X (2015) Interesterification of soybean oil and methyl stearate catalyzed by guanidine-functionalized SBA-15 silica. J Am Oil Chem Soc 92(6):915–925. https://doi.org/10.1007/s11746-015-2651-2

20. Viriato RLS, de Queirós M, da Gama MAS, Ribeiro APB, Gigante ML (2018) Milk fat as a structuring agent of plastic lipid bases. Food Res Int 111:120–129. https://doi.org/10.1016/j.foodres.2018.05.015

21. Guo Y et al (2020) Synthesis, physicochemical properties, and health aspects of structured lipids: a review. Compr Rev Food Sci Food Saf 19(2):759–800. https://doi.org/10.1111/1541-4337.12537

22. Moreira DKT et al (2020) Synthesis and characterization of structured lipid rich in behenic acid by enzymatic interesterification. Food Bioprod Process 122:303–310. https://doi.org/10.1016/j.fbp.2020.06.005

23. Singh PK, Chopra R, Garg M, Dhiman A, Dhyani A (2022) Enzymatic interesterification of vegetable oil: a review on physicochemical and functional properties, and its health effects. J Oleo Sci 71(12):ess22118. https://doi.org/10.5650/jos.ess22118

24. Moreira DKT, Santos PS, Gambero A, Macedo GA (2017) Evaluation of structured lipids with behenic acid in the prevention of obesity. Food Res Int 95:52–58. https://doi.org/10.1016/j.foodres.2017.03.005

25. Zhao S-Q et al (2014) Characteristics and feasibility of trans- free plastic fats through lipozyme tl im-catalyzed interesterification of palm stearin and akebia trifoliata variety australis seed oil. J Agric Food Chem 62(14):3293–3300. https://doi.org/10.1021/jf500267e

26. Willis WM, Lencki RW, Marangoni AG (1998) Lipid modification strategies in the production of nutritionally functional fats and oils. Crit Rev Food Sci Nutr 38(8):639–674. https://doi.org/10.1080/10408699891274336

27. Norizzah AR, Nur Azimah K, Zaliha O (2018) Influence of enzymatic and chemical interesterification on crystallisation properties of refined, bleached and deodourised (RBD) palm oil and RBD palm kernel oil blends. Food Res Int 106:982–991. https://doi.org/10.1016/j.foodres.2018.02.001

28. Gibon V, Kellens M (2014) Latest developments in chemical and enzymatic interesterification for commodity oils and specialty fats. In: Trans fats replacement solutions, Elsevier, pp 153–185. https://doi.org/10.1016/B978-0-9830791-5-6.50013-7

29. Zhang L et al (2015) Effect of lard quality on chemical interesterification catalyzed by KOH/glycerol. J Am Oil Chem Soc 92(4):513–521. https://doi.org/10.1007/s11746-015-2619-2

30. Santoro V et al (2018) Formation of by-products during chemical interesterification of lipids: detection and characterization of dialkyl ketones by non-aqueous reversed-phase liquid chromatography-high resolution mass spectrometry and gas chromatography-mass spectrometry. J Chromatogr A 1581–1582:63–70. https://doi.org/10.1016/j.chroma.2018.11.001

31. Ferreira-Dias S, Osório N, Tecelão C (2002) Bioprocess technologies for production of structured lipids as nutraceuticals. In: Current developments in biotechnology and bioengineering, Elsevier, pp 209–237. https://doi.org/10.1016/B978-0-12-823506-5.00007-2

32. Ahmadi L, Wright AJ, Marangoni AG (2008) Chemical and enzymatic interesterification of tristearin/ triolein-rich blends: chemical composition, solid fat content and thermal properties. Eur J Lipid Sci Technol 110(11):1014–1024. https://doi.org/10.1002/ejlt.200800058

33. Ribeiro APB, Grimaldi R, Gioielli LA, Gonçalves LAG (2009) Zero trans fats from soybean oil and fully hydrogenated soybean oil: physico-chemical properties and food applications. Food Res Int 42(3):401–410. https://doi.org/10.1016/j.foodres.2009.01.012

34. da Silva RC et al (2010) Structured lipids obtained by chemical interesterification of olive oil and palm stearin. LWT Food Sci Technol 43(5):752–758. https://doi.org/10.1016/j.lwt.2009.12.010

35. Adhikari P et al (2010) Production of *trans* -free margarine stock by enzymatic interesterification of rice bran oil, palm stearin and coconut oil. J Sci Food Agric 90(4):703–711. https://doi.org/10.1002/jsfa.3872

36. Guedes AMM, Ming CC, Ribeiro APB, da Silva RC, Gioielli LA, Gonçalves LAG (2014) Physicochemical properties of interesterified blends of fully hydrogenated *Crambe abyssinica*

oil and soybean oil. J Am Oil Chem Soc 91(1):111–123. https://doi.org/10.1007/s11746-013-2360-7

37. Lopes TIB, Ribeiro MDMM, Ming CC, Grimaldi R, Gonçalves LAG, Marsaioli AJ (2016) Comparison of the regiospecific distribution from triacylglycerols after chemical and enzymatic interesterification of high oleic sunflower oil and fully hydrogenated high oleic sunflower oil blend by carbon-13 nuclear magnetic resonance. Food Chem 212:641–647. https://doi.org/10.1016/j.foodchem.2016.06.024

38. Hu P, Xu X, Yu LL (2017) Interesterified trans-free fats rich in sn-2 nervonic acid prepared using Acer truncatum oil, palm stearin and palm kernel oil, and their physicochemical properties. LWT Food Sci Technol 76:156–163. https://doi.org/10.1016/j.lwt.2016.10.054

39. Zhang Z, Ye J, Lee WJ, Akoh CC, Li A, Wang Y (2021) Modification of palm-based oil blend via interesterification: physicochemical properties, crystallization behaviors and oxidative stabilities. Food Chem 347:129070. https://doi.org/10.1016/j.foodchem.2021.129070

40. Zuin JC, de P. Gandra RL, Ribeiro APB, Ract JNR, Macedo JA, Macedo GA (2022) Comparing chemical and enzymatic synthesis of rich behenic lipids products: technological and nutritional potential. Food Sci Technol https://doi.org/10.1590/fst.105821

41. De Greyt W, Dijkstra Albert J (2008) Trans fatty acids. Wiley. https://doi.org/10.1002/9780470697658

42. Rousseau D, Ghazani SM, Marangoni AG (2017) Extraction and analysis of lipids. In: Food lipids, Fourth edition. |CRC Press, Boca Raton, Taylor & Francis, pp 149–184. https://doi.org/10.1201/9781315151854-13

43. Zhou H, Zhang Z, Lee WJ, Xie X, Li A, Wang Y (2021) Acyl migration occurrence of palm olein during interesterification catalyzed by sn-1,3 specific lipase. LWT 142:111023. https://doi.org/10.1016/j.lwt.2021.111023

44. Zhang Z, Lee WJ, Wang Y (2021) Evaluation of enzymatic interesterification in structured triacylglycerols preparation: a concise review and prospect. Crit Rev Food Sci Nutr 61(19):3145–3159. https://doi.org/10.1080/10408398.2020.1793725

45. Dijkstra AJ (2015) Interesterification, chemical or enzymatic catalysis. Lipid Technol 27(6):134–136. https://doi.org/10.1002/lite.201500029

46. Ballestra D Produce trans-free fats. Science behind technology. www.desmetballestra.com

47. Ferreira-Dias S, Osório NM, Rodrigues J, Tecelão C (2019) Structured lipids for foods. In: Encyclopedia of food chemistry, Elsevier, pp 357–369. https://doi.org/10.1016/B978-0-08-100596-5.21766-6

48. Fernandez-Lafuente R (2010) Lipase from Thermomyces lanuginosus: Uses and prospects as an industrial biocatalyst. J Mol Catal B Enzym 62(3–4):197–212. https://doi.org/10.1016/j.molcatb.2009.11.010

49. Cortez DV, de Castro HF, Andrade GSS (2016) Potencial catalítico de lipases ligadas ao micélio de fungos filamentosos em processos de biotransformação. Quim Nova. https://doi.org/10.21577/0100-4042.20160163

50. Sturt NRM, Vieira SS, Moura FCC (2019) Catalytic activity of sulfated niobium oxide for oleic acid esterification. J Environ Chem Eng 7(1):102866. https://doi.org/10.1016/j.jece.2018.102866

51. Tecelão C, Perrier V, Dubreucq E, Ferreira-Dias S (2019) Production of human milk fat substitutes by interesterification of tripalmitin with ethyl oleate catalyzed by candida parapsilosis lipase/acyltransferase. J Am Oil Chem Soc 96(7):777–787. https://doi.org/10.1002/aocs.12250

52. Mota DA et al (2020) Production of low-calorie structured lipids from spent coffee grounds or olive pomace crude oils catalyzed by immobilized lipase in magnetic nanoparticles. Bioresour Technol 307:123223. https://doi.org/10.1016/j.biortech.2020.123223

53. Zhang J et al (2020) Heterogeneous catalytic esterification of oleic acid under sub/supercritical methanol over γ-Al2O3. Fuel 268:117359. https://doi.org/10.1016/j.fuel.2020.117359

54. Akil E, da S. Pereira A, El-Bacha T, Amaral PFF, Torres AG (2020) Efficient production of bioactive structured lipids by fast acidolysis catalyzed by Yarrowia lipolytica lipase, free and immobilized in chitosan-alginate beads, in solvent-free medium. Int J Biol Macromol 163:910–918. https://doi.org/10.1016/j.ijbiomac.2020.06.282

55. Khodadadi M, Aziz S, St-Louis R, Kermasha S (2013) Lipase-catalyzed synthesis and characterization of flaxseed oil-based structured lipids. J Funct Foods 5(1):424–433. https://doi.org/10.1016/j.jff.2012.11.015

56. Kavadia MR, Yadav MG, Vadgama RN, Odaneth AA, Lali AM (2019) Production of trans-free interesterified fat using indigenously immobilized lipase. Prep Biochem Biotechnol 49(5):444–452. https://doi.org/10.1080/10826068.2019.1566142

57. Jaeger K (1994) Bacterial lipases. FEMS Microbiol Rev 15(1):29–63. https://doi.org/10.1016/0168-6445(94)90025-6

58. Reyes HR, Hill CG (1994) Kinetic modeling of interesterification reactions catalyzed by immobilized lipase. Biotechnol Bioeng 43(2):171–182. https://doi.org/10.1002/bit.260430211

59. Wong DW (1995) Food enzymes: structure and mechanism. Springer Science & Business Media

60. Islam MdN, Zhang M, Adhikari B (2014) The inactivation of enzymes by ultrasound—a review of potential mechanisms. Food Rev Intl 30(1):1–21. https://doi.org/10.1080/87559129.2013.853772

61. Wang Z, Liu L, Liu L, Liu T, Li C, Sun L (2019) 1,3-Dioleoyl-2-palmitoylglycerol-rich triacylglycerol characterization by three processing methods. Int J Food Prop 22(1):1156–1171. https://doi.org/10.1080/10942912.2019.1632345

62. Yue K, Yang H, Li J, Bi Y, Zhang L, Lou W-Y (2023) UPU structured lipids and their preparation methods: a mini review. Food Biosci 55:103009. https://doi.org/10.1016/j.fbio.2023.103009

63. Zhao J et al (2022) Green synthesis of polydopamine functionalized magnetic mesoporous biochar for lipase immobilization and its application in interesterification for novel structured lipids production. Food Chem 379:132148. https://doi.org/10.1016/j.foodchem.2022.132148

Chapter 2
Lipases as Biocatalysts for Enzymatic Interesterification

2.1 Introduction

The biocatalysts used for enzymatic interesterification reactions are lipases, which come from a heterogeneous class of enzymes obtained from fungi, yeasts, and bacteria of animal and/or plant origin [1]. The preference for microbial-derived lipases over those of animal or plant origin is typically due to several factors. Most microorganisms secrete extracellular enzymes, have rapid growth rates, are easier to genetically manipulate, and produce enzymes that are more stable and active [2–4].

One of their main characteristics is to be active at the oil/water interface [5]. The catalytic mechanism of lipases, commonly known as interfacial activation, determines a broad range of actions when in contact with hydrophobic environments, whether substrates or supports and the structure of most lipases include a polypeptide chain with a hydrophobic and hydrophilic face and the lipases also have a kind of "entry gate," also called a "lid" which controls the enzyme's active site, generally, when lipases are in a hydrophobic environment, their active site is exposed and interacts with hydrophobic areas surrounding the active center, isolating it from the rest of the reaction medium, in an aqueous environment, lipase is unstable and tends to be in the closed form (inactive site) [6, 7]. Therefore, lipases are efficient in the hydrolysis of oils and fats and/or in non-aqueous environments. Figure 2.1 presents the general equilibrium mechanism between open and closed conformations.

As can be observed in Fig. 2.1, for lipase catalysis to occur, interfacial activation is necessary beforehand. However, it is worth noting that depending on the type and source of the enzyme, along with the production conditions, some lipases may differ in the characteristics of the "lid," with reports of lipases having up to two "lids" or no "lid," characterized as atypical lipases as they do not require interfacial activation [8–11].

Lipases can be characterized as regioselective enzymes, originating from *Aspergillus niger* and *Rhizopus arrhizus*, displaying high selectivity toward positions *sn*-1,3; however, they do not act on esters in position *sn*-2 due to steric hindrance. On

V. Alves et al., *Chemical and Enzymatic Interesterification for Food Lipid Production*, Chemistry of Foods, https://doi.org/10.1007/978-3-031-67405-1_2

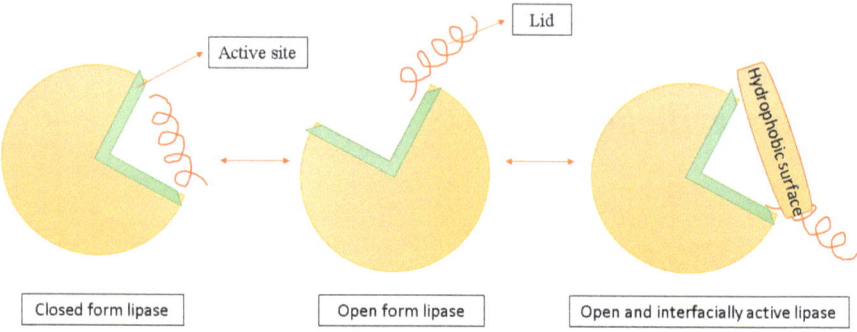

Fig. 2.1 Mechanism of interfacial activation of lipases. Figure adapted from Ref. [7]

the other hand, *Candida parapsilosis* lipase is regioselective for the *sn*-2 position, hydrolyzing the acyl group of the FA at this position more rapidly than at *sn*-1 or *sn*-3, under specific conditions; stereo-selective lipases hydrolyze the bonds located at *sn*-1 and *sn*-3 positions at different rates, they can also be selective/specific, acting particularly on a certain class of FAs, regardless of their position, examples include *Pseudomonas fluorescens* and *Humicola lanuginosa*, which show specificity for *sn*-1, and *Candida antarctica* B. for *sn*-3 [12]. Additionally, some lipases can act in a non-specific manner, such as *Candida cylindracea* and *Staphylococcus aureus*, these cleave FAs at all three positions of the TAG without distinction, producing products similar to chemical interesterification [13].

2.2 Lipases

These lipases can be used either in the form of free (non-immobilized, i.e., in their soluble form) lipases or immobilized on solid supports (adhered to different matrices), allowing for reuse in continuous or batch bioreactors, this approach helps reduce the costs associated with the reaction process [14]. Enzyme immobilization is the process of attaching enzymes to suitable substrates, which can be achieved through various techniques, fixation to a carrier/support including physical adsorption or covalent bonding; entrapment by inclusion or into a membrane network (encapsulation); support-free enzyme aggregation as entrapment, and chemical crosslinking and physical immobilization [15]. One of the commonly employed methods is adsorption, which is reversible and involves the use of organic or inorganic carriers, such as chitosan, silica, ion exchange resin, among others [16]. Immobilized enzymes exhibit enhanced functionality as they can be easily recovered and reused in continuous processes, in other words, they can be extracted and separated from the product through physical processes such as filtration, allowing for straightforward handling in large-scale industrial applications and additionally, immobilized enzymes offer more efficient enzymatic stability at specific temperatures and pH levels when compared to free lipases [1, 15, 17, 18]. Thus, the main objectives of enzyme immobilization are: to make enzymes insoluble in water, increase operational stability, facilitate reuse, enhance process control through product/enzyme separation and product formation, automate processes, especially for designing reactors for continuous processes, and minimize allosteric enzyme inhibition [19].

For the production of commercial lipases, fungal sources are primarily used due to their versatility and greater stability compared to plant or animal sources [17]. In general, commercial lipases used for the production of SLs are commonly in immobilized form, such as Lipozyme RM IM derived from the lipase of *Rhizomucor miehei*, immobilized on a macroporous anion exchange resin support, specifically targeting *sn*-1,3 positions; Lipozyme TL IM, derived from the lipase of *Thermomyces lanuginosus*, is immobilized on silica gel and specifically targets *sn*-1,3 positions and Novozymes 435 is derived from the lipase of *Candida antarctica* B. immobilized on a macroporous acrylic resin, it exhibits specificity for *sn*-1,3 positions or nonspecificity depending on the type of substrate used in the reaction [20]. Examples of immobilized enzymes used in the production of SLs are presented in Table 2.1.

2.3 Factors Influencing Enzymatic Interesterification

Even with the use of specific lipases (regioselective and/or stereo-selective), undesirable and inevitable reaction processes can still occur, resulting in the excessive formation of by-products (free FA, MAGs, and DAGs). To prevent this, it is necessary to mitigate acyl migration by controlling certain parameters such as temperature, biocatalyst concentration, type of immobilization support, water content with moisture control, solvent type, pH and acidity, reaction time, and reaction system [20, 33].

Immobilized enzymes are more stable and can operate at higher temperatures compared to non-immobilized enzymes, with a commonly used reaction temperature of 70 °C; however, it is worth noting that each type of enzyme requires different optimal temperatures for its catalytic activity [17, 34, 35]. One of the main factors influencing the cost of industrial production is the concentration of immobilized enzyme used in interesterification, generally, concentrations of 3 to 10% (w/v) are used. Enzymatic reactions also require low moisture in the reaction system; if the water content exceeds 1%, it can cause excessive hydrolysis, while a water content below 0.01% may hinder the total hydration of the lipase and reduce the initial hydrolysis rate [17, 34, 35]. Solvent systems can also interfere with enzymatic interesterification since one of the main characteristics of lipases is their activity at the oil/water interface and the solvent polarity or partition coefficient interferes with the specificity and activation of the lipases' active site, with the preferable to carry out interesterification with oils and fats without solvents in the reaction medium, using immobilized enzymes in the synthesis due to their greater stability [5, 33, 36, 37].

Table 2.1 Immobilized lipases used in the production of structured lipids

Syntheses and biocatalysts	Sources of lipid	Results	References
Acidolysis + Lipozyme TL IM and Lipozyme RM IM	Emu oil + ω3—EPA	SL developed was able to randomly incorporate EPA into sn-2,3 positions by RM IM, while TL IM mainly incorporated EPA in sn-1,3	[21]
Acidolysis + Lipozyme RM IM	Basa catfish oil + sesame oil	SL showed potential as a substitute for HMF in infant formulas	[22]
Enzymatic interesterification + Lipozyme RM IM	Palm stearin + rice bran oil	SL proved suitable for use and/or substitution as trans-fat-free margarine fat	[23]
Enzymatic interesterification + Lipozyme TL IM and *Rhizopus sp.*	Olive oil + soybean oil and crambe hardfat rich in behenic acid	C57Bl/6 mice showed lower weight gain when consuming SLs containing behenic acid, attributed to the increased lipid excretion in feces, without toxicity or diarrhea	[24]
Enzymatic interesterification + Lipozyme RM IM	Palm olein + high oleic sunflower oil + behenic acid	Satisfactory physical and rheological properties at room temperature for application in bakery products	[25]
Acidolysis + organic solvent + Novozymes 435, Lipozyme 435, TL IM e RM IM	Microalgae oils from *Nannochloropsis oculata* and *Isochrysis galbana*	The analyzed lipases emerged as promising biocatalysts for the production of HMF substitutes from microalgae oils	[26]
Enzymatic interesterification + Lipozyme TL IM	Coconut and high oleic sunflower oils	The developed SLs showed promise in the application of more translucent edible films, and interesterification increased oleic acid in the sn-2 position	[27]
Enzymatic interesterification + Lipozyme RM IM	*Cinnamomum camphora* seed oil and fully hydrogenated palm oil	SL proved to be suitable for use and/or substitution of cocoa butter, providing a low-cost edible alternative	[28]
Acidolysis + enzymatic interesterification + Lipozyme TL IM and *Rhizopus oryzae*	Spent coffee grounds or olive pomace crude oils + capric acid	*Rhizopus oryzae* proved to be a promising biocatalyst for the production of SL with higher activity than Lipozyme TL IM in both oils, with a preference for acidolysis	[14]

(continued)

Table 2.1 (continued)

Syntheses and biocatalysts	Sources of lipid	Results	References
Enzymatic interesterification + Novozymes 435	Peanut oil + coconut oil	The body weight and serum triglyceride levels in C57BL/6 J mice in the DAG group with coconut oil were lower than with peanut oil	[29]
Enzymatic interesterification + Lipozyme RM IM	Soybean oil + coconut oil	Anti-obesity effect in C57BL/6 J mice with lower body mass gain and visceral fat accumulation	[30]
Enzymatic interesterification + Lipozyme TL IM	Olive oil + soybean oil and crambe hardfat rich in behenic acid	Interesterification increased the saturated FA content in the sn-1,3 position by 47%. The SLs exhibited suitable plastic fat properties for replacing fats in food	[31]
Acidolysis + Novozymes 435 + Lipozyme RM IM + Lipozyme 435 and Lipozyme TL IM	Single-cell oil from *Schizochytrium sp.* + caprylic acid	SL with caprylic acid in sn-1,3 and DHA in sn-2. SLs may promote better DHA absorption in infants, with potential for infant formulas	[32]

2.4 Alternative Vegetable Lipases to Immobilized Lipases

Although enzymatic interesterification catalyzed by lipase appears to be a more suitable pathway when compared to acidolysis or chemical methods, as it has fewer limitations in the recovery of SLs, commercial immobilized lipases have high costs and may bring some limitations, including potential reduction in enzymatic activity after immobilization, variations in reaction kinetics, and mass transfer issues [38]. Therefore, the use of non-commercial enzymes and/or more economically viable immobilization supports and techniques are alternatives to be explored to reduce production costs related to the biocatalyst, an example is vegetable lipases, which exhibit broad versatility and stability in organic environments [33].

In this context, some studies propose the use of non-commercial lipases such as those from *R. oryzae* immobilized on magnetite nanoparticles and *Carica papaya* self-immobilized in papaya latex as an alternative to commercial lipases. These non-commercial lipases were faster and preferable, showing a yield of up to 69% for SLs compared to Lipozyme TL IM with a yield of 50% for SLs in oils from coffee grounds, olive pomace, and capric acid [39]. There are also reports of the use of *Y. lipolytica* lipase microencapsulated by ionotropic gelation with alginate and chitosan, which showed higher thermal stability compared to commercial immobilized lipases [40]. And the use of *Euphorbia characias* as a plant biocatalyst, a species native to the Mediterranean [41]. These data bring efficient approaches with the potential to expand the applications of plant lipases as biocatalysts for the production of SLs. It is worth noting that the use of agro-forestry and industrial residues as immobilization supports can also be an alternative to reduce the costs of immobilized lipases [33]. Cai et al. [42] developed new carriers to immobilize lipases through covalent bonds using magnetic microspheres applied with polysaccharides. The immobilized lipases in the study retained 85% of the initial catalytic activity after nine enzyme reuse syntheses and showed better thermal stability compared to free lipases and the commercial enzyme Lipozyme RM IM.

2.5 Enzymatic Conditioning

Immobilized enzymes may need to undergo a conditioning step before the interesterification reaction, following the manufacturer's instructions. This conditioning is performed to remove air (deaeration) and excess moisture (drying) from the enzyme particles, preventing the oxidation of the interesterified oil and limiting losses due to hydrolysis [43]. Typically, conditioning is carried out in a batch system under vacuum and agitation, using the reaction oil or a refined oil with low acidity for a certain period, usually 30 min. The oil used in conditioning will contain significant amounts of free FAs and water. Therefore, when the conditioning time is complete, the oil must be replaced with new oil in a batch washing system. The conditioning progress is monitored by measuring the free FA index through titration [44] after each batch until a constant acidity is achieved, indicating that the enzyme is ready for the interesterification reaction.

References

1. Soumanou MM, Pérignon M, Villeneuve P (2013) Lipase-catalyzed interesterification reactions for human milk fat substitutes production: a review. Eur J Lipid Sci Technol 115(3):270–285. https://doi.org/10.1002/ejlt.201200084
2. Bharathi D, Rajalakshmi G (2019) Microbial lipases: an overview of screening, production and purification. Biocatal Agric Biotechnol 22:101368. https://doi.org/10.1016/j.bcab.2019.101368
3. Javed S et al (2018) Bacterial lipases: a review on purification and characterization. Prog Biophys Mol Biol 132:23–34. https://doi.org/10.1016/j.pbiomolbio.2017.07.014
4. Liu S, Li Z, Yu B, Wang S, Shen Y, Cong H (2020) Recent advances on protein separation and purification methods. Adv Colloid Interface Sci 284:102254. https://doi.org/10.1016/j.cis.2020.102254
5. Schmid RD, Verger R (1998) Lipases: interfacial enzymes with attractive applications. Angew Chem Int Ed 37(12):1608–1633. https://doi.org/10.1002/(SICI)1521-3773(19980703)37:12%3c1608::AID-ANIE1608%3e3.0.CO;2-V
6. Brzozowski AM et al (1991) A model for interfacial activation in lipases from the structure of a fungal lipase-inhibitor complex. Nature 351(6326):491–494. https://doi.org/10.1038/351491a0
7. Rodrigues RC et al (2019) Immobilization of lipases on hydrophobic supports: immobilization mechanism, advantages, problems, and solutions. Biotechnol Adv 37(5):746–770. https://doi.org/10.1016/j.biotechadv.2019.04.003

8. Mulinari J, Oliveira JV, Hotza D (2020) Lipase immobilization on ceramic supports: An overview on techniques and materials. Biotechnol Adv 42:107581. https://doi.org/10.1016/j.biotechadv.2020.107581

9. Adlercreutz P (2013) Immobilisation and application of lipases in organic media. Chem Soc Rev 42(15):6406. https://doi.org/10.1039/c3cs35446f

10. Yaacob N, Mohamad Ali MS, Salleh AB, Rahman RNZRA, Leow ATC (2016) Toluene promotes lid 2 interfacial activation of cold active solvent tolerant lipase from Pseudomonas fluorescens strain AMS8. J Mol Graph Model 68:224–235. https://doi.org/10.1016/j.jmgm.2016.07.003

11. Remonatto D, Miotti RH Jr, Monti R, Bassan JC, de Paula AV (2022) Applications of immobilized lipases in enzymatic reactors: a review. Process Biochem 114:1–20. https://doi.org/10.1016/j.procbio.2022.01.004

12. Kadhum AAH, Shamma MN (2017) Edible lipids modification processes: a review. Crit Rev Food Sci Nutr 57(1):48–58. https://doi.org/10.1080/10408398.2013.848834

13. Ferreira ML, Tonetto GM (2017) Enzymatic synthesis of structured triglycerides. Springer International Publishing, Cham. https://doi.org/10.1007/978-3-319-51574-8

14. Mota DA et al (2020) Production of low-calorie structured lipids from spent coffee grounds or olive pomace crude oils catalyzed by immobilized lipase in magnetic nanoparticles. Bioresour Technol 307:123223. https://doi.org/10.1016/j.biortech.2020.123223

15. Guo Y et al (2020) Synthesis, physicochemical properties, and health aspects of structured lipids: a review. Compr Rev Food Sci Food Saf 19(2):759–800. https://doi.org/10.1111/1541-4337.12537

16. Sirisha VL, Jain A, Jain A (2016) Enzyme immobilization. pp 179–211. https://doi.org/10.1016/bs.afnr.2016.07.004

17. Utama QD, Sitanggang AB, Adawiyah DR, Hariyadi P (2019) Lipase-catalyzed interesterification for the synthesis of medium-long-medium (MLM) structured lipids. Food Technol Biotechnol 57(3):305–318. https://doi.org/10.17113/ftb.57.03.19.6025

18. Liu Y, WeiZhuo X, Wei X (2022) A review on lipase-catalyzed synthesis of geranyl esters as flavor additives for food, pharmaceutical and cosmetic applications. Food Chem Adv 1:100052. https://doi.org/10.1016/j.focha.2022.100052

19. Sheldon RA, Basso A, Brady D (2021) New frontiers in enzyme immobilisation: robust biocatalysts for a circular bio-based economy. Chem Soc Rev 50(10):5850–5862. https://doi.org/10.1039/D1CS00015B

20. Kim BH, Akoh CC (2015) Recent research trends on the enzymatic synthesis of structured lipids. J Food Sci 80(8). https://doi.org/10.1111/1750-3841.12953

21. Akanbi TO, Barrow CJ (2015) Lipase-catalysed incorporation of EPA into emu oil: formation and characterisation of new structured lipids. J Funct Foods 19:801–809. https://doi.org/10.1016/j.jff.2014.11.010

22. Zou X, Jin Q, Guo Z, Xu X, Wang X (2016) Preparation of human milk fat substitutes from basa catfish oil: combination of enzymatic acidolysis and modeled blending. Eur J Lipid Sci Technol 118(11):1702–1711. https://doi.org/10.1002/ejlt.201500591

23. Ornla-ied P, Sonwai S, Lertthirasuntorn S (2016) Trans-free margarine fat produced using enzymatic interesterification of rice bran oil and hard palm stearin. Food Sci Biotechnol 25(3):673–680. https://doi.org/10.1007/s10068-016-0118-3

24. Moreira DKT, Santos PS, Gambero A, Macedo GA (2017) Evaluation of structured lipids with behenic acid in the prevention of obesity. Food Res Int 95:52–58. https://doi.org/10.1016/j.foodres.2017.03.005

25. Kok W-M, Chuah C-H, Cheng S-F (2017) Enzymatic synthesis of structured lipids with behenic acid at the sn-1, 3 positions of triacylglycerols. Food Sci Biotechnol. https://doi.org/10.1007/s10068-017-0271-3

26. He Y, Qiu C, Guo Z, Huang J, Wang M, Chen B (2017) Production of new human milk fat substitutes by enzymatic acidolysis of microalgae oils from Nannochloropsis oculata and Isochrysis galbana. Bioresour Technol 238:129–138. https://doi.org/10.1016/j.biortech.2017.04.041

27. Moore MA, Akoh CC (2017) Enzymatic interesterification of coconut and high oleic sunflower oils for edible film application. J Am Oil Chem Soc 94(4):567–576. https://doi.org/10.1007/s11746-017-2969-z

28. Ma X, Hu Z, Mao J, Xu Y, Zhu X, Xiong H (2019) Synthesis of cocoa butter substitutes from Cinnamomum camphora seed oil and fully hydrogenated palm oil by enzymatic interesterification. J Food Sci Technol 56(2):835–845. https://doi.org/10.1007/s13197-018-3543-x

29. Lu H, Guo T, Fan Y, Deng Z, Luo T, Li H (2020) Effects of diacylglycerol and triacylglycerol from peanut oil and coconut oil on lipid metabolism in mice. J Food Sci 85(6):1907–1914. https://doi.org/10.1111/1750-3841.15159

30. Ji S, Xu F, Zhang N, Wu Y, Ju X, Wang L (2021) Dietary a novel structured lipid synthesized by soybean oil and coconut oil alter fatty acid metabolism in C57BL/6J mice. Food Biosci 44:101396. https://doi.org/10.1016/j.fbio.2021.101396

31. Zuin JC, de P. Gandra RL, Ribeiro APB, Ract JNR, Macedo JA, Macedo GA (2022) Comparing chemical and enzymatic synthesis of rich behenic lipids products: technological and nutritional potential. Food Sci Technol. https://doi.org/10.1590/fst.105821

32. Zou X et al (2023) Bioimprinted lipase-catalyzed synthesis of medium- and long-chain structured lipids rich in docosahexaenoic acid for infant formula. Food Chem 424:136450. https://doi.org/10.1016/j.foodchem.2023.136450

33. Ferreira-Dias S, Osório NM, Rodrigues J, Tecelão C (2019) Structured lipids for foods. In: Encyclopedia of food chemistry, Elsevier, pp 357–369. https://doi.org/10.1016/B978-0-08-100596-5.21766-6

34. Basso A, Serban S (2019) Industrial applications of immobilized enzymes—a review. Mol Cataly 479:110607. https://doi.org/10.1016/j.mcat.2019.110607

35. Maruyama T, Nakajima M, Uchikawa S, Nabetani H, Furusaki S, Seki M (2000) Oil-water interfacial activation of lipase for interesterification of triglyceride and fatty acid. J Am Oil Chem Soc 77(11). https://doi.org/10.1007/s11746-000-0176-4

36. Choong TSY et al (2018) Kinetic study of lipase-catalyzed glycerolysis of palm olein using Lipozyme TLIM in solvent-free system. PLoS ONE 13(2):e0192375. https://doi.org/10.1371/journal.pone.0192375

37. Palacios D, Ortega N, Rubio-Rodríguez N, Busto MD (2019) Lipase-catalyzed glycerolysis of anchovy oil in a solvent-free system: Simultaneous optimization of monoacylglycerol synthesis and end-product oxidative stability. Food Chem 271:372–379. https://doi.org/10.1016/j.foodchem.2018.07.184

38. DiCosimo R, McAuliffe J, Poulose AJ, Bohlmann G (2013) Industrial use of immobilized enzymes. Chem Soc Rev 42(15):6437. https://doi.org/10.1039/c3cs35506c

39. Costa CM et al (2018) Production of MLM type structured lipids from grapeseed oil catalyzed by non-commercial lipases. Euro J Lipid Sci Technol 120(1):1700320. https://doi.org/10.1002/ejlt.201700320

40. Akil E, da S Pereira A, El-Bacha T, Amaral PFF, Torres AG (2020) Efficient production of bioactive structured lipids by fast acidolysis catalyzed by Yarrowia lipolytica lipase, free and immobilized in chitosan-alginate beads, in solvent-free medium. Int J Biol Macromol 163:910–918. https://doi.org/10.1016/j.ijbiomac.2020.06.282

41. Villeneuve P (2003) Plant lipases and their applications in oils and fats modification. Eur J Lipid Sci Technol 105(6):308–317. https://doi.org/10.1002/ejlt.200390061

42. Cai Z et al (2019) Lipase immobilized on layer-by-layer polysaccharide-coated Fe3O4@ SiO2 microspheres as a reusable biocatalyst for the production of structured lipids. ACS Sustain Chem Eng 7(7):6685–6695

43. Gibon V, Kellens M (2014) Latest developments in chemical and enzymatic interesterification for commodity oils and specialty fats. In: Trans fats replacement solutions, Elsevier, pp 153–185. https://doi.org/10.1016/B978-0-9830791-5-6.50013-7

44. AOCS—American Oil Chemists' Society (2009) Official methods and recommended practices of the American Oil Chemists' Society, 6th edn. AOCS Press, Champaign, IL, Chicago

Chapter 3
Flow Bioreactors for the Biocatalytic Process of Enzymatic Interesterification

3.1 Introduction

Enzymatic synthesis through biocatalysis by immobilized enzymes promotes safe, sustainable, and efficient processes, as mentioned earlier. These processes are carried out in reactors, which provide specific reaction conditions regarding flow patterns and/or operating modes to ensure an adequate enzyme–substrate interaction for the production of biocompounds [1]. So, the choice of the biocatalysis operating mode should be evaluated according to the characteristics and needs of the reaction process.

The reactors used can be fluidized-bed reactors, packed-bed reactors, and continuous and batch stirred tank reactors, in solvent-free or organic solvent-containing conditions [2]. Because these reactors allow the use of different configurations and operating methods, they are commonly employed in heterogeneous enzymatic processes such as batch, fed-batch, and continuous systems [3].

The continuous flow reactor stands out for its ease of scaling up and low shear stress, preventing enzyme desorption, and is considered more productive when compared to the batch reactor; however, at the laboratory scale, batch stirred tank reactors have been currently used because they offer greater flexibility and simplicity in the process they are more suitable for producing smaller volumes of products and are a good option to prevent support particles from being damaged by shear forces (tank and impeller) [4, 5].

The main types of enzymatic processes are carried out in batch or continuous reactors, as shown in the scheme of Fig. 3.1.

© The Author(s), under exclusive license to Springer Nature Switzerland AG 2024
V. Alves et al., *Chemical and Enzymatic Interesterification for Food Lipid Production*,
Chemistry of Foods, https://doi.org/10.1007/978-3-031-67405-1_3

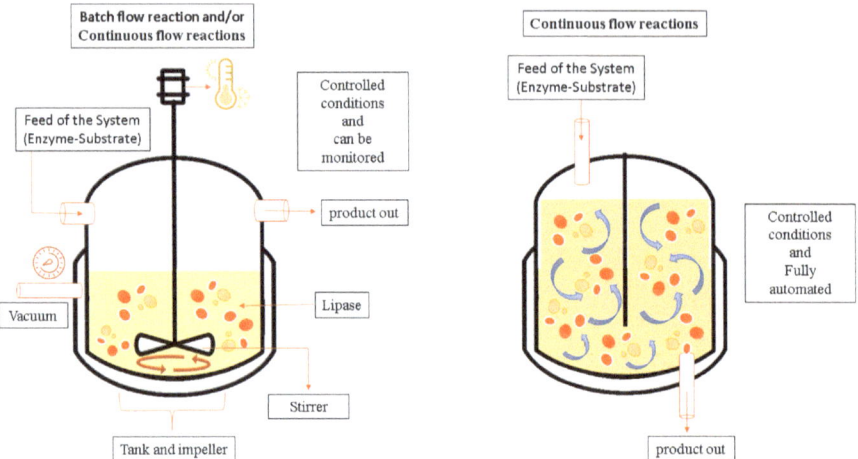

Fig. 3.1 Schematic representation of the main types of reactors used in enzymatic syntheses. Figure adapted from Ref. [6]

3.2 Synthesis Process in the Reactor

3.2.1 Batch Flow Reaction

The synthesis process in the reactor begins with the addition of the lipid substrate to the enzyme, where they remain dispersed in the lipid substrate, under mechanical agitation in a closed system under specific and controlled conditions (agitation, temperature, pressure, oxygen, and vacuum can be monitored). In this type of reaction, the stirred tank reactor is commonly employed, as it offers greater simplicity in operation, equipped with agitation and temperature control [7]. At the end of the reaction, the final product (SLs) is separated from the enzyme through recovery by filtration [8].

Batch reaction is associated with several advantages such as greater dispersion of the enzyme in the lipid substrate, low operational cost, simplicity of structure, and flexibility for application in various types of operations, high conversion factors of final products, easy control of reaction kinetics, and other process data [9]. However, it is worth noting that this type of reaction is commonly used on a laboratory scale, as mentioned earlier, due to the low production volume, gradual loss of enzymatic activity through successive reuses, need for longer reaction time to maintain the enzymatic conversion rate, cleaning and reloading of the reactor between batches, need for more labor, and variation in the final product that occurs between batches, making it difficult for large-scale application [7, 10].

In the literature, there is an incidence of the use of batch reactors, with agitation parameters between 200 and 350 rpm at temperatures of 60–75 °C [11–15].

3.2.2 Continuous Flow Reactions

Reactors used in continuous processes can be fluidized bed reactors, packed bed reactors, and/or continuous stirred tank reactors, with the choice related to the characteristics and needs of each reaction process.

In continuous reactions, the biocatalyst and the substrate flow through the reactors, and the reaction time is determined by the flow rate and reactor volume, characterized by the residence time, which is usually long, especially for immobilized lipases, due to lower productivity loss [16].

The advantages associated with this process include greater control of reagent mixing, lower cost in optimizing conditions, higher energy efficiency, and easier product recovery [17]. When compared to the batch system, continuous reaction in interesterification can be faster and more economically viable. Generally, when used in the industry, continuous reactors are fully automated to ensure rigorous control and standardization of process conditions [18].

3.2.3 Other Reactor Types

Biocatalyses with lipases are also carried out in stirred tank reactors, fluidized bed reactors, biocatalytic membrane reactors, bubble column reactors, and vortex flow reactors [6].

Stirred tank reactors consist basically of a tank containing a mechanical agitator, which is used to improve the efficiency of the reaction blend's homogenization [19]. These reactors are widely used because they can operate in batch, fed-batch, and continuous modes. Batch stirred tank reactors are equipped with temperature control devices and an agitation system, and no material is removed until the reaction is stopped, allowing for monitoring and control [20]. Continuous stirred tank reactors also feature temperature and agitation control systems, and the feeding and withdrawal of reaction substrates and products are carried out continuously, enhancing their capacity in large-scale industrial applications [21].

The fluidized bed reactor system consists of particles (immobilized enzymes) kept in constant motion, also known as fluidization, as they pump fluid through the bed [21]. This type of system is considered a hybrid of stirred tank and packed-bed reactors [22]. It's regarded as an easy-to-operate system because it provides high mass transfer, generated by turbulence induced by the feed flow rate, which is sufficiently high to elevate the catalyst particles, yielding good results in continuous mode with immobilized enzymes [23].

Bubble column reactors are generally applied in the production of bioproducts such as enzymes, proteins, and antibiotics but are gaining prominence in enzymatic synthesis [24], particularly in esterification reactions catalyzed by free lipases (non-immobilized) [25]. These reactors are multiphase, meaning the beds can contain a liquid phase or a solid–liquid suspension, and they consist of a cylindrical vessel with

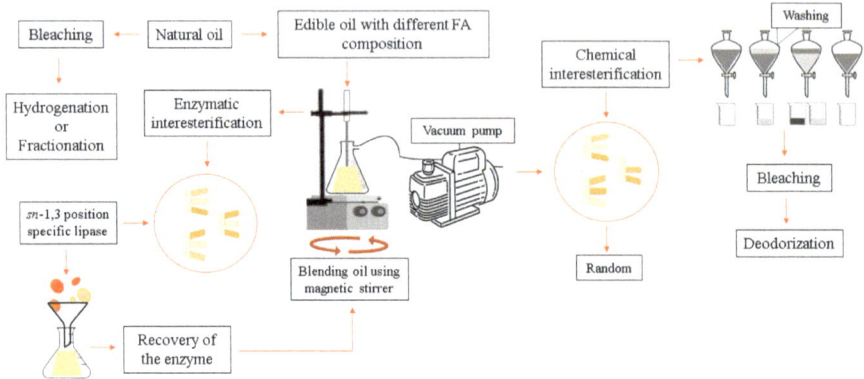

Fig. 3.2 Schematic representation of the mechanism of lipid modification processes

a gas distributor at the bottom, which is injected through the reaction bed, agitating the medium [26].

In membrane biocatalytic reactors, continuous, batch, and fed-batch modes can be used. These reactors feature membranes that serve multiple functions, acting as separators for the recovery of reaction products, physical barriers for catalyst recovery, or supports for enzyme immobilization, for example [27]. Membrane reactors are commonly employed in biphasic liquid systems due to the membranes' role in fluid segregation [28, 29]. The advantages associated with this type of reactor include the fact that catalysis and product separation occur in the same unit due to the presence of membranes; however, limitations include reduced enzymatic activity over time and membrane fouling, which occurs due to complex interactions between membranes and feed solutions, thereby reducing performance and lifespan, leading to increased operational costs, so, it is essential to choose suitable membrane materials for these reactors [30–34].

Vortex flow reactors consist of a rotating cylinder inside a stationary external cylinder (concentric cylinders) and resemble fluidized bed reactors, with the difference that the rotation speed of the inner cylinder can be changed independently of the axial speed to keep the biocatalyst particles suspended, facilitating reactor operation at low flow rates, which is a limitation for fluidized bed reactors [35]. Advantages include lower shear stress, which is desirable when using immobilized lipases on sensitive support immobilization particles [36]. This reactor can still be operated in batch, continuous, or semi-continuous mode, and its application is determined by the characteristics of Taylor–Couette flow, which depends on the reactor configuration [37].

The mechanism of lipid modification processes, including the interesterification reaction, is schematically represented in Fig. 3.2.

References

1. Erickson LE (2011) Bioreactors for commodity products. In: Comprehensive biotechnology, Elsevier, pp 653–658. https://doi.org/10.1016/B978-0-08-088504-9.00236-1
2. Spier MR, Vandenberghe L, Medeiros APB, Soccol C (2011) Application of different types of bioreactors in bioprocesses. In: Bioreactors: design, properties and applications, pp 53–87
3. de Meneses AC et al (2019) Benzyl butyrate esterification mediated by immobilized lipases: evaluation of batch and fed-batch reactors to overcome lipase-acid deactivation. Process Biochem 78:50–57. https://doi.org/10.1016/j.procbio.2018.12.029
4. Tamborini L, Fernandes P, Paradisi F, Molinari F (2018) Flow bioreactors as complementary tools for biocatalytic process intensification. Trends Biotechnol 36(1):73–88. https://doi.org/10.1016/j.tibtech.2017.09.005
5. Ferreira-Dias S, Osório NM, Rodrigues J, Tecelão C (2019) Structured lipids for foods. In: Encyclopedia of food chemistry, Elsevier, pp 357–369. https://doi.org/10.1016/B978-0-08-100596-5.21766-6
6. Remonatto D, Miotti RH Jr, Monti R, Bassan JC, de Paula AV (2022) Applications of immobilized lipases in enzymatic reactors: a review. Process Biochem 114:1–20. https://doi.org/10.1016/j.procbio.2022.01.004
7. Christopher LP, Hemanathan Kumar H, Zambare VP (2014) Enzymatic biodiesel: challenges and opportunities. Appl Energy 119:497–520. https://doi.org/10.1016/j.apenergy.2014.01.017
8. SenSharma S, Kumar G, Sarkar A (2023) Immobilized enzyme reactors for bioremediation. In: Metagenomics to bioremediation, Elsevier, pp 641–657. https://doi.org/10.1016/B978-0-323-96113-4.00021-4.
9. Tan T, Lu J, Nie K, Deng L, Wang F (2010) Biodiesel production with immobilized lipase: a review. Biotechnol Adv 28(5):628–634. https://doi.org/10.1016/j.biotechadv.2010.05.012
10. Nielsen PM, Brask J, Fjerbaek L (2008) Enzymatic biodiesel production: technical and economical considerations. Eur J Lipid Sci Technol 110(8):692–700. https://doi.org/10.1002/ejlt.200800064
11. Speranza P, Ribeiro APB, Macedo GA (2016) Application of lipases to regiospecific interesterification of exotic oils from an Amazonian area. J Biotechnol 218:13–20. https://doi.org/10.1016/j.jbiotec.2015.11.025
12. Norizzah AR, Nur Azimah K, Zaliha O (2018) Influence of enzymatic and chemical interesterification on crystallisation properties of refined, bleached and deodourised (RBD) palm oil and RBD palm kernel oil blends. Food Res Int 106:982–991. https://doi.org/10.1016/j.foodres.2018.02.001
13. Moreira DKT et al (2020) Synthesis and characterization of structured lipid rich in behenic acid by enzymatic interesterification. Food Bioprod Process 122:303–310. https://doi.org/10.1016/j.fbp.2020.06.005
14. Ji S, Xu F, Zhang N, Wu Y, Ju X, Wang L (2021) Dietary a novel structured lipid synthesized by soybean oil and coconut oil alter fatty acid metabolism in C57BL/6J mice. Food Biosci 44:101396. https://doi.org/10.1016/j.fbio.2021.101396
15. Zhou H, Zhang Z, Lee WJ, Xie X, Li A, Wang Y (2021) Acyl migration occurrence of palm olein during interesterification catalyzed by sn-1,3 specific lipase. LWT 142:111023. https://doi.org/10.1016/j.lwt.2021.111023
16. de Souza ROMA, Miranda LSM (2014) Continuous flow reactions: from green chemistry towards a green process. Revista Virtual de Química 6(1). https://doi.org/10.5935/1984-6835.20140004
17. Itabaiana I, de Mariz e Miranda LS, de Souza ROMA (2013) Towards a continuous flow environment for lipase-catalyzed reactions. J Mol Catal B Enzym 85–86:1–9. https://doi.org/10.1016/j.molcatb.2012.08.008
18. Wang X, Liu X, Zhao C, Ding Y, Xu P (2011) Biodiesel production in packed-bed reactors using lipase–nanoparticle biocomposite. Bioresour Technol 102(10):6352–6355. https://doi.org/10.1016/j.biortech.2011.03.003

19. Paula AV, Nunes GFM, Osório NM, Santos JC, de Castro HF, Ferreira-Dias S (2015) Continuous enzymatic interesterification of milkfat with soybean oil produces a highly spreadable product rich in polyunsaturated fatty acids. Eur J Lipid Sci Technol 117(5):608–619. https://doi.org/10.1002/ejlt.201400316

20. Balcão VM, Paiva AL, Xavier Malcata F (1996) Bioreactors with immobilized lipases: state of the art. Enzyme Microb Technol 18(6):392–416. https://doi.org/10.1016/0141-0229(95)00125-5

21. Aguieiras ECG, Cavalcanti-Oliveira ED, Freire DMG (2015) Current status and new developments of biodiesel production using fungal lipases. Fuel 159:52–67. https://doi.org/10.1016/j.fuel.2015.06.064

22. Poppe JK, Fernandez-Lafuente R, Rodrigues RC, Ayub MAZ (2015) Enzymatic reactors for biodiesel synthesis: present status and future prospects. Biotechnol Adv 33(5):511–525. https://doi.org/10.1016/j.biotechadv.2015.01.011

23. Fidalgo WRR, Ceron A, Freitas L, Santos JC, de Castro HF (2016) A fluidized bed reactor as an approach to enzymatic biodiesel production in a process with simultaneous glycerol removal. J Ind Eng Chem 38:217–223. https://doi.org/10.1016/j.jiec.2016.05.005

24. Yu H, Lee M-W, Shin H, Park K-M, Chang P-S (2019) Lipase-catalyzed solvent-free synthesis of erythorbyl laurate in a gas-solid-liquid multiphase system. Food Chem 271:445–449. https://doi.org/10.1016/j.foodchem.2018.07.134

25. Hansen RB, Agerbaek MA, Nielsen PM, Rancke-Madsen A, Woodley JM (2020) Esterification using a liquid lipase to remove residual free fatty acids in biodiesel. Process Biochem 97:213–221. https://doi.org/10.1016/j.procbio.2020.06.005

26. Kantarci N, Borak F, Ulgen KO (2005) Bubble column reactors. Process Biochem 40(7):2263–2283. https://doi.org/10.1016/j.procbio.2004.10.004

27. Stankiewicz AI, Moulijn JA Process intensification: transforming chemical engineering. Chem Eng Prog. https://www.aiche.org/sites/default/files/docs/news/010022_cep_stankiewicz.pdf

28. Aghababaie M, Beheshti M, Razmjou A, Bordbar A-K (2019) Two phase enzymatic membrane reactor for the production of biodiesel from crude Eruca sativa oil. Renew Energy 140:104–110. https://doi.org/10.1016/j.renene.2019.03.069

29. Heyse A, Plikat C, Ansorge-Schumacher M, Drews A (2019) Continuous two-phase biocatalysis using water-in-oil Pickering emulsions in a membrane reactor: evaluation of different nanoparticles. Catal Today 331:60–67. https://doi.org/10.1016/j.cattod.2017.11.032

30. Brunetti A, Zito PF, Giorno L, Drioli E, Barbieri G (2018) Membrane reactors for low temperature applications: an overview. Chem Eng Process Process Intensificat 124:282–307. https://doi.org/10.1016/j.cep.2017.05.002

31. Saleh J, Tremblay AY, Dubé MA (2010) Glycerol removal from biodiesel using membrane separation technology. Fuel 89(9):2260–2266. https://doi.org/10.1016/j.fuel.2010.04.025

32. Sokač T, Gojun M, Tušek AJ, Šalić A, Zelić B (2020) Purification of biodiesel produced by lipase catalysed transesterification by ultrafiltration: selection of membranes and analysis of membrane blocking mechanisms. Renew Energy 159:642–651. https://doi.org/10.1016/j.renene.2020.05.132

33. Jochems P, Satyawali Y, Diels L, Dejonghe W (2011) Enzyme immobilization on/in polymeric membranes: status, challenges and perspectives in biocatalytic membrane reactors (BMRs). Green Chem 13(7):1609. https://doi.org/10.1039/c1gc15178a

34. Rios GM, Belleville MP, Paolucci D, Sanchez J (2004) Progress in enzymatic membrane reactors—a review. J Memb Sci 242(1–2):189–196. https://doi.org/10.1016/j.memsci.2003.06.004

35. Ibáñez-González MJ, Cooney CL (2007) Studies on protein adsorption in a vortex flow reactor. Process Biochem 42(12):1592–1601. https://doi.org/10.1016/j.procbio.2007.08.012

36. Giordano RLC, Giordano RC, Cooney CL (2000) Performance of a continuous Taylor–Couette–Poiseuille vortex flow enzymic reactor with suspended particles. Process Biochem 35(10):1093–1101. https://doi.org/10.1016/S0032-9592(00)00143-6

37. Resende MM, Vieira PG, Sousa R Jr, Giordano RLC, Giordano RC (2004) Estimation of mass transfer parameters in a Taylor-Couette-Poiseuille heterogeneous reactor. Braz J Chem Eng 21(2):175–184. https://doi.org/10.1590/S0104-66322004000200006

Chapter 4
Applications of Structured Lipids in Foods

4.1 Human Milk Fat (HMF) Substitutes

TAGs of HMF have unsaturated FAs in the *sn*-1,3 positions of glycerol, with saturated FAs in the *sn*-2 position, mainly palmitic acid. This TAG structure is crucial for the efficient absorption of palmitic acid as palmitoyl-monoglyceride [1]. In infant formulas, the use of TAGs from vegetable oils and cow's milk can lead to the formation of insoluble calcium soaps with saturated FAs, which are released through the action of pancreatic lipase that is regioselective for *sn*-1,3, causing inefficient absorption of calcium and FAs by infants [2, 3].

In the market, there are already established substitutes for HMF in infant formulas, such as Betapol, developed by IOI Loders Croklaan (Netherlands), which is produced through enzymatic acidolysis of fractionated palm oil with FAs from high oleic sunflower oil. Another substitute is InFat, marketed by Advanced Lipids, which is a structured TAG with a maximum total palmitic acid composition of 38% and a minimum of 60% palmitic acid in the *sn*-2 position [1].

The enzymatic syntheses used for HMF substitutes involve either acidolysis or interesterification with ethyl esters using regioselective *sn*-1,3 lipases as catalysts, only a source of TAG rich in *sn*-2 palmitic acid (tripalmitin) is required, which can come from sources such as palm stearin, palm oil, fractionated palm stearin, butterfat, lard, or other animal milk fat, for example (Fig. 4.1) [1, 4]. Other oil sources, such as microalgae and fish oil and their fractions rich in EPA, DHA, and arachidonic acid (AA), are also used along with blends of different vegetable oils such as olive oil, sunflower oil, soybean oil, hazelnut oil, among others, to produce analogs of HMF [5–10].

It is worth noting that the majority of studies found in the literature on substitutes for HMF use palm oil fractions as a source of palmitic acid and microalgae oil and fish oil as a source of long-chain unsaturated FAs [8–12]. However, despite the associated

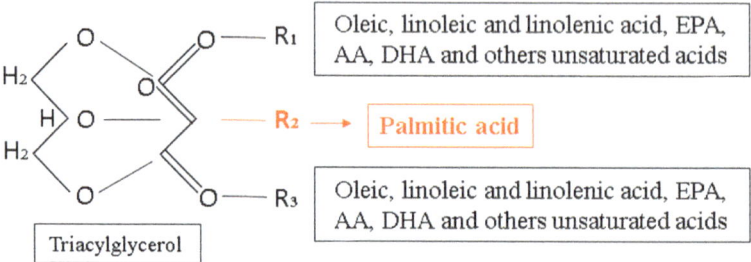

Fig. 4.1 Diagram of the structured lipid structure for a human milk fat substitute. Adapted figure from Ref. [4]

benefits of AA and DHA, these FAs are highly susceptible to oxidative deterioration, which, according to [13], can be minimized through the microencapsulation of medium- and long-chain TAGs containing AA.

4.2 Plastic Fats

The term plastic fats refer to semi-solid fats primarily used in margarines and baking applications. Traditionally, these fats were obtained through the process of partial hydrogenation. However, due to the prohibition of trans-fats and their association with health issues and NCDs [14], interesterification has been employed as an alternative. Interesterification helps maintain the composition of unsaturated FAs without forming trans-isomers [15]. The reaction can be carried out through enzymatic interesterification with free or immobilized lipases, as well as through chemical interesterification.

Interesterification imparts desirable characteristics to plastic fats, especially for margarines and hardstocks; these characteristics include plasticity and spreadability between 4 and 10 °C; a melting point higher than room temperature, with melting occurring above 35 °C to avoid a waxy organoleptic sensation, also results in a crystalline morphology of the β' polymorph type, providing a smooth texture, shiny surface, and good melting properties [15–17]. Blends of the hardfats of the palm stearin, palm kernel, palm oil, fully hydrogenated coconut oil and crambe oil, among others, are commonly used with vegetable oils rich in MUFAs and PUFAs with canola oil, sunflower oil, soybean oil, cottonseed oil, for example. Some commercially available examples include NovaLipid™ and SansTrans™, composed of naturally stable oils and fats with enhanced characteristics for margarines, vegetable fats, and customized blends. Another product line is Crokvitol™ (Crokvitol™ Stand, Crokvitol™ Allround, Crokvitol™ Vitality) by Loders Croklaan, which provides an alternative to hydrogenated palm oil and plastic fats with low or zero trans-fats for margarines, baking products, and spreads [1].

Several studies available in the literature report promising results regarding the application of interesterification for the production of margarines and trans-fat-free fats from different lipid sources [15, 18–21].

4.3 Equivalents and Substitutes for Cocoa Butter

Cocoa butter (CB), for example, is an essential ingredient in the production of chocolates and their derivatives, constitutes the main fat component ($\approx$25–36% of the final product) obtained through the pressing of cocoa mass (*Theobroma cacao*) and CB imparts texture and characteristic melting behavior, snap, hardness, demolding contraction, shine, and shape to the continuous phase of the chocolate and also acts as a dispersion matrix for cocoa solids and sugar [22]. This is due to its specific physical and chemical properties, being semi-solid at room temperature ($\approx$20 to 25 °C) and liquid at body temperature ($\approx$37 °C). It is composed of three molecules of FAs bound to glycerol, with the main components being stearic acid—S (33.3–40.2%), oleic acid—O (32.7–37.0%), and palmitic acid—P (24.1–33.7%) and its TAGs composition includes POP (13.8–21.8%), POS (26.3–44.8%), and SOS (20.0–29.4%) [23].

Despite the high demand for CB, its production has been relatively decreasing due to climate and environmental changes affecting cocoa plantations, this has significantly contributed to the increase in its price compared to most vegetable oils and fats [24]. In this context, as an alternative to this issue, blends of CB from different sources have been used to partially or fully replace CB, these fats are referred to as cocoa butter alternatives (CBA) [25, 26]. The CBAs can be referred to as equivalents to CB, derived from non-lauric vegetable fats, while substitutes for CB come from lauric vegetable fats, both can be blended with CB without altering its properties; however, CB equivalents possess physical and chemical properties similar to CB, whereas CB substitutes are chemically different from CB [25].

The lipid source commonly used to produce CB equivalents and substitutes is palm oil and its fractions, this is due to its easy accessibility, low cost, and TAG composition, which exhibit desirable physicochemical characteristics after interesterification, such as the formation of β' crystals [27]. In the literature, a series of studies are available addressing the production of CB equivalents and substitutes through interesterification and acidolysis [22, 24, 25, 28, 29].

4.4 Low-Calorie Structured Lipids

Low-calorie lipids are characterized by a structure where short- and/or medium-chain FAs are present at *sn*-1,3 positions, and long-chain FAs are at *sn*-2. The low caloric value (approximately 5 kcal/g) is attributed to the fact that medium-chain FAs are metabolized more rapidly compared to long-chain FAs, this is because they are absorbed through the intestinal wall without resynthesis into triglycerides in intestinal cells, avoiding accumulation in adipose cells due to frequent β-oxidation, in other words, they have direct transport to the liver and not to the lymphatic system [30–32]. Thus, they provide a quick source of energy, making them suitable for individuals with lipid metabolic disorders, such as malabsorption [1]. The initial food applications of low-calorie SLs were through enteral and parenteral nutrition, followed by various clinical settings, including thrombosis prevention, nitrogen balance, and immune functions [33]. These applications demonstrated more efficient rates of lipid absorption as a quick energy source compared to other diets offered [34, 35].

Long-chain saturated FAs, such as behenic acid, are also interesting as low-calorie lipids due to their anti-obesogenic effect, they exhibit low absorption and inhibit pancreatic lipase activity, leading to high release and low absorption of triglycerides in the gastrointestinal tract and depending on their dietary composition and restructuring within the triglyceride, this partial absorption may increase HDL-C levels, decrease LDL-C levels, and reduce blood triglycerides [36, 37].

References [37, 38] obtained and characterized SLs rich in behenic acid from olive oil, soybean oil, and fully hydrogenated crambe oil with an anti-obesity effect through the inhibition of pancreatic lipase [31]. When evaluating the anti-obesity effects through in vivo methods, it is observed that rats consuming SLs containing behenic acid showed lower weight gain. This was attributed to a significant increase in lipid excretion in feces, without any signs of toxicity or diarrhea. This demonstrated sufficient absorption of essential FAs in the gastrointestinal tract. Silva et al. [39] demonstrated that these SLs inhibited pancreatic lipase by up to 20% in an in vitro digestion model with rats. This resulted in improved blood glucose levels, reductions in non-alcoholic fatty liver disease markers, and changes in microbiota composition. Gandra et al. [40] produced nanoemulsions with these SLs and observed good stability during storage. After the in vitro gastric digestion phase, there were no signs of coalescence. In cellular studies, there was no reduction in viability, and when the samples were added to LPS-stimulated macrophages (Raw 264.7), there was a decrease in inflammatory cytokines (IL-6 and TNF-α).

The production of low-calorie SLs can be achieved through acidolysis and inter-esterification reactions, involving medium-chain FAs (such as caprylic or capric acids) with different lipid sources such as olive oil, soybean oil, sesame oil, avocado oil, grape seed oil, pumpkin seed oil, pine nut oil, for example [41–47].

There are several commercial medications used for long-term obesity control, considered low-calorie due to their reduction of lipid absorption in the gastrointestinal tract. Among them, we highlight some that incorporate FAs into their composition. Orlistat is commercially known as Xenical® and is one of the most widely used chemical compounds in the treatment of obesity. It is a semi-synthetic derivative of lipstatin, an extremely lipophilic lactone with amphiphilic specificities. Orlistat actively functions in the gastrointestinal tract by selectively inhibiting pancreatic and gastric lipases, partially inhibiting the hydrolysis of TAGs, and subsequently reducing the absorption of MAGs and free FAs [48]. Olestra, also commercially known as Olean®, is an acylated sucrose polyester obtained by esterifying long-chain FAs and methyl esters of vegetable oils, such as soybean, corn, sunflower, among others. It is not hydrolyzed by gastric or pancreatic enzymes due to its large size and number of nonpolar fats, and consequently, it has no caloric or toxic value, however, it raises nutritional concerns as it is indigestible and passes directly through the gastrointestinal tract, resulting in the excretion of lipids in the feces and additionally, it interferes with the absorption of some lipophilic components, such as fat-soluble vitamins, carotenoids, and certain phytochemicals [43]. Caprenin (Caprenin®) is a SL that contains FAs such as caprylic, capric, and behenic acids esterified to a glycerol fraction. As behenic acid is only partially absorbed, caprylic and capric acids are metabolized more rapidly than other long-chain FAs, Caprenin provides only 5 kcal/g, making it considered a low-calorie SL [33, 49]. Neobee is another reduced-calorie fat composed of fractionated caprylic and capric acids derived from coconut or palm kernel oils. It provides 6.8 kcal/g and includes different products, for instance, Neobec 1053 and Neobee M-5 contain caprylic and capric acids, while Neobee 1095 is composed solely of capric acid [50].

4.5 Structured Lipids Enriched with Essential Fatty Acids

The foundation of the Western diet is characterized by the excessive consumption of saturated fats, salt, and sugar, leading to various health issues such as obesity and metabolic syndrome, it can worsen and/or trigger existing or pre-existing conditions related to NCDs [51, 52], Thus, there is a need for effective dietary approaches that act non-aggressively in controlling these issues. The interesterification reaction, through the restructuring and incorporation of new FAs into TAG, allows the development of SLs with added nutritional value and specific health effects, which can include essential FAs in their composition.

The FAs of greatest interest to incorporate into SLs are: linoleic acid (C18:2, omega-6), alpha-linolenic acid (C18:3, omega-3), AA (C20:4; omega-6), EPA (C20:5, omega-3), DHA (C22:6, omega-3), gamma-linolenic acid (GLA), and conjugated linoleic acid (CLA) [1, 53]. However, it is worth noting that SLs enriched with essential FAs are not only available to provide new sources of PUFAs but also aim to optimize their concentration and increase their availability and metabolic absorption through their incorporation into everyday Western diet food products, such as vegetable oils, margarines, and bakery products, for example [54], given that EPA and DHA are commonly found in marine fish [1], which are not commonly included in the Western diet.

The synthesis of SLs rich in long-chain PUFAs can be carried out through the reactions of acidolysis and interesterification, as reported in studies available in the literature [55–60].

4.6   Other Types of Structured Fats for Applications in Food

4.6.1   Structured Phospholipids

Just as with SLs, the main reason for modifying phospholipids (PLs) is for specific applications, providing technical or physiological properties that their native form does not exhibit. PLs offer more advantages compared to PUFA TAGs because they are incorporated into lipoproteins in the bloodstream and thus are more efficient carriers to various tissues and organs, including blood cells such as platelets and erythrocytes [61].

The molecular structure of PLs can be altered by enzymatic or chemical means, with enzymatic processes allowing for the synthesis of new PL compounds for functional and nutritional applications, such as *sn*-1 lysophospholipids (*sn*-1 Lyso-PLs); *sn*-2 lysophospholipids (*sn*-2 Lyso-PLs); *sn*-1 modified phosphatidylcholine (PC); and *sn*-2 modified PC, which cannot be obtained through chemical routes [1]. These modifications in PLs can be carried out in organic medium and/or solvent-free by acidolysis with FFAs or interesterification between PLs and FA esters [1].

Lysophospholipids (Lyso-PLs) are obtained by partial hydrolysis of the PL molecule with the removal of one FA. When the hydrolysis of PLs is catalyzed by phospholipases A2, *sn*-2 Lyso-PL is obtained; and consequently, if phospholipases A1 or regioselective sn-1,3 lipases are used, the final product is *sn*-1 Lyso-PL [62, 63]. Lyso-PLs can also be synthesized by alcoholysis, esterification of glycerophosphorylcholine, and thermodynamic transacylation of 2-acyl lysophospholipids into 1-acyl lysophospholipids [62].

Refrences [64–66] PLs were modified through enzymatic catalysis to incorporate saturated FAs, medium-chain FAs, and/or PUFAs, and it was found that the absorption of these FAs in the human body improved when incorporated into the PLs.

4.6.2   Monoacylglycerols and Diacylglycerols

MAGs and DAGs have been used as important emulsifiers in the food and pharmaceutical industries [67]. Regarding health benefits, partial acylglycerols have a positive effect on elevated plasma TAG levels and anti-obesogenic action. Additionally, *sn*-2 MAGs have beneficial effects on human health because they are more easily absorbed by the intestinal mucosa and are directly used for the resynthesis of new TAGs [68].

Refrences [69, 70] discuss that *sn*-2 MAGs are involved in the synthesis and degradation of endocannabinoids, which can regulate appetite, pain sensation, inflammation, and lipid metabolism. Eom et al. [71], evaluated the anti-obesogenic effects in an in vivo study with mice on a diet rich in synthesized DAGs, and observed a reduction in body weight gain and plasma biochemical markers of obesity (total cholesterol, triacylglycerol, and glucose levels) in these animals compared to a diet rich in TAGs.

The catalysis generally employed for the industrial production of MAGs and DAGs has been chemical, requiring the use of expensive and toxic solvents. However, this catalysis can be sustainably performed through lipase-catalyzed glycerolysis of a TAG, fat, or oil in a non-aqueous medium [72, 73]. The yield of MAGs or DAGs obtained through enzymatic catalysis is determined by the type of lipase used as biocatalysts (non-selective or *sn*-1,3 regioselective lipase). That is, with a *sn*-1,3 regioselective lipase, *sn*-2 monoacylglycerols (*sn*-2-MAGs) or *sn*-1,3-DAGs are obtained; using a non-regioselective lipase, all types of MAG and DAG can be obtained [1]. For the preparation of *sn*-1,3-DAGs, some studies have utilized glycerolysis of vegetable oils, as well as direct esterification of glycerol with FFAs catalyzed by commercial immobilized lipases [74–76].

DAGs produced from unsaturated vegetable oils are easily oxidized; however, DAGs produced with high melting point vegetable oils, such as palm oil and/or its derivatives, exhibit greater resistance, higher melting points, and improved plasticity. They can be used to substitute hardfats [77]. The primary application of DAGs has been in baking, mainly due to their amphiphilicity, lower interfacial tension between oil and water, and ease of emulsification. Although it is used as a food additive or crystallization inhibitor, there is limited data on the use of DAGs as a primary lipid base [78]. Most research on the use of DAGs focuses on improving the synthesis process as well as green alternatives, aiming to enhance process optimization or discussing their physiological functions. However, studies exploring the physical and chemical properties of oils enriched with DAGs, as well as the impact of processing conditions on the molecular structure of DAGs, are still limited [79–82].

4.6.3 Esterified Propoxylated Glycerol (EPG)

It's an alternative structuring fat consisting of acetyl epoxides and polyols, formed by the conjugation of glycerol and propylene oxide under alkaline conditions, followed by esterification with FA (its propoxy group is linked between the glycerol structure and the FA) [77]. According to [83], EPG is non-toxic to humans (when not exceeding appropriate intake levels), and there are no reports of accumulation in the human body after consumption.

EPG exhibits a two-phase state, meaning it demonstrates rheological properties of liquid oil and solid texture properties, with pleasant organoleptic characteristics. This two-phase state depends on the average degree of propoxylation (the number of propylene glycol oligomers between the glycerol structure and the FA) and the composition of FAs, where a higher degree of propoxylation results in a more liquid EPG [83].

In metabolic terms, TAG presents 2-MAG as an intermediate metabolite, which is converted into TAG chylomicrons, increasing the metabolic load [84]. Meanwhile, the propylene glycol group that exists between the glycerol structure and the acyl

FAs in EPG is separated from the FFAs after digestion [77]. The methyl group blocks lipase access to the ester bond, leading to no formation of 1-MAG, thus preventing the formation of chylomicron TAG. Consequently, EPGs with a high degree of propoxylation are less sensitive to lipases and may have more beneficial effects on human metabolism [83].

Propoxylglycerol esterified, based on vegetable oils, can replace 50–85% of fat in most applications, mainly used for calorie reduction. It can be applied in reduced-calorie chocolate bars and baked goods, as well as in other applications such as snacks (potato chips, corn snacks, and chicken nuggets), plant-based protein products, beverages, coffee, tea, and dairy analogs, promise as a new lipid substitute [77].

4.6.4 Powdered Oil

The edible oils are typically susceptible to oxidation during storage, leading to rapid oxidative degradation due to exposure to light, heat, moisture, oxygen, etc. Another concern associated with the challenge of maintaining vegetable oils stable for long periods is the choice and selection of packaging materials [85]. For this reason, edible oils can be converted into powder through the encapsulation process using various wall materials as well as processing conditions [86].

The microencapsulated oil, commonly known as powdered oil, involves stabilizing the emulsion through emulsification or structuring techniques (mixing hydrophilic and lipophilic phases), where the aqueous phase can be a protein, polysaccharide, or a different component that binds and adsorbs to the outer layer of lipid molecules [87]. This is followed by encapsulation and formation of dry powders, which can include oven drying, freeze-drying, fluidized bed agglomeration, spray drying, nanospray drying, freeze-drying, and electrohydrodynamic process [87–93].

The wall material should be water-soluble and can be made from vegetable oil, corn syrup, high-quality protein, stabilizer, and emulsifier [77]. It is worth noting that the choice of wall material is also a major concern among consumers, as a portion of the population prefers vegan or vegetarian foods, and a plant-based source of wall material would be ideal [94].

Refrence [95] evaluated the oxidative stability of nanoencapsulated powdered oil (soybean and mustard) through freeze-drying based on accelerated shelf-life tests at 60 °C for 24 days. The authors noted that the process had a positive impact on maintaining the oil quality during storage. Srivastava and Mishra [96] reported the protective effect of edible vegetable oil powder (sunflower and sesame) through microencapsulation via the spray drying process. The authors suggested that bioactive ingredients as wall materials tend to enhance thermo-oxidative stability and minimize the energy requirement during handling and storage. According to [97], nanoemulsion-based encapsulated powdered oil provides better oxidative stability compared to free-form vegetable oils.

Powdered oils can be added to a variety of instant food matrices, such as powdered soup, cake mix, and powdered milk, etc., and are increasingly used as ingredients in tea, coffee, smoothies, cookies, yogurt, cake mix, and powdered soups [86]. Another interesting application is the development of nutraceutical delivery vehicles where the oil can be encapsulated for targeted release of bioactive compounds according to its FA composition. That is, under specific conditions, the outer wall of the encapsulated oil ruptures, and the oil can be released at specific locations in the gastrointestinal tract (GIT) [98].

Currently, powdered oils are the main component in formulating ketogenic diets for weight loss, obtained from coconut, flaxseed, chia seeds, and fish. Their FAs provide a feeling of satiety, thus reducing excess caloric intake, while also supplying essential FAs like omega-3 [86].

4.6.5   Inert Gas Spray Oil

Refrence [77] discusses the use of edible oil spray, which, unlike liquid edible vegetable oil, features a design that allows the oil to be evenly distributed, thus helping to control and reduce dietary intake. The edible oil spray employs a binary packaging design, which is a multiprotection material packaging system consisting of valves, multilayer vacuum bags, and aluminum cans that are explosion-proof and fireproof, preventing oil leakage. Due to the opacity of the system and its good sealing performance, exposure of the product to air and light is reduced, making it less susceptible to oxidation. The authors emphasize that to improve oxidative stability, an inert gas pressure container can be proposed as an alternative to microencapsulation, allowing for improved atomization efficiency and enhancing the sealing effect of the oil spray, as well as preventing exposure to oxygen and prolonging shelf-life.

References

1. Ferreira-Dias S, Osório NM, Rodrigues J, Tecelão C (2019) Structured lipids for foods. In: Encyclopedia of food chemistry, Elsevier, pp 357–369. https://doi.org/10.1016/B978-0-08-100 596-5.21766-6
2. López-López A et al (2001) The influence of dietary palmitic acid triacylglyceride position on the fatty acid, calcium and magnesium contents of at term newborn faeces. Early Hum Dev 65:S83–S94. https://doi.org/10.1016/S0378-3782(01)00210-9
3. Akanbi TO, Adcock JL, Barrow CJ (2013) Selective concentration of EPA and DHA using thermomyces lanuginosus lipase is due to fatty acid selectivity and not regioselectivity. Food Chem 138(1):615–620. https://doi.org/10.1016/j.foodchem.2012.11.007
4. Yue K, Yang H, Li J, Bi Y, Zhang L, Lou W-Y (2023) UPU structured lipids and their preparation methods: a mini review. Food Biosci 55:103009. https://doi.org/10.1016/j.fbio.2023.103009
5. Turan D, Şahin Yeşilçubuk N, Akoh CC (2012) Production of human milk fat analogue containing docosahexaenoic and arachidonic acids. J Agric Food Chem 60(17):4402–4407. https://doi.org/10.1021/jf3012272

 6. Soumanou MM, Pérignon M, Villeneuve P (2013) Lipase-catalyzed interesterification reactions for human milk fat substitutes production: a review. Eur J Lipid Sci Technol 115(3):270–285. https://doi.org/10.1002/ejlt.201200084
 7. Ferreira-Dias S, Tecelão C (2014) Human milk fat substitutes: advances and constraints of enzyme-catalyzed production. Lipid Technol 26(8):183–186. https://doi.org/10.1002/lite.201400043
 8. Ghosh M, Sengupta A, Bhattacharyya DK, Ghosh M (2016) Preparation of human milk fat analogue by enzymatic interesterification reaction using palm stearin and fish oil. J Food Sci Technol 53(4):2017–2024. https://doi.org/10.1007/s13197-016-2180-5
 9. He Y, Qiu C, Guo Z, Huang J, Wang M, Chen B (2017) Production of new human milk fat substitutes by enzymatic acidolysis of microalgae oils from Nannochloropsis oculata and Isochrysis galbana. Bioresour Technol 238:129–138. https://doi.org/10.1016/j.biortech.2017.04.041
10. Wei W, Jin Q, Wang X (2019) Human milk fat substitutes: past achievements and current trends. Prog Lipid Res 74:69–86. https://doi.org/10.1016/j.plipres.2019.02.001
11. Nagachinta S, Akoh CC (2013) Synthesis of structured lipid enriched with omega fatty acids and sn-2 palmitic acid by enzymatic esterification and its incorporation in powdered infant formula. J Agric Food Chem 61(18):4455–4463. https://doi.org/10.1021/jf400634w
12. Wang J et al (2016) Selective synthesis of human milk fat-style structured triglycerides from microalgal oil in a microfluidic reactor packed with immobilized lipase. Bioresour Technol 220:132–141. https://doi.org/10.1016/j.biortech.2016.08.023
13. Korma SA et al (2019) Spray-dried novel structured lipids enriched with medium-and long-chain triacylglycerols encapsulated with different wall materials: characterization and stability. Food Res Int 116:538–547. https://doi.org/10.1016/j.foodres.2018.08.071
14. Viriato RLS, de Queirós M, da Gama MSA, Ribeiro APB, Gigante ML (2018) Milk fat as a structuring agent of plastic lipid bases. Food Res Int 111:120–129. https://doi.org/10.1016/j.foodres.2018.05.015
15. Li Y, Zhao J, Xie X, Zhang Z, Zhang N, Wang Y (2018) A low trans margarine fat analog to beef tallow for healthier formulations: optimization of enzymatic interesterification using soybean oil and fully hydrogenated palm oil. Food Chem 255:405–413. https://doi.org/10.1016/j.foodchem.2018.02.086
16. Danthine S, Lefébure E, Trinh HN, Blecker C (2014) Effect of palm oil enzymatic interesterification on physicochemical and structural properties of mixed fat blends. J Am Oil Chem Soc 91(9):1477–1487. https://doi.org/10.1007/s11746-014-2494-2
17. deMan JM, Finley JW, Hurst WJ, Lee CY (2018) Principles of food chemistry. Springer International Publishing, Cham. https://doi.org/10.1007/978-3-319-63607-8
18. De Martini Soares FAS, Osório NM, da Silva RC, Gioielli LA, Ferreira-Dias S (2013) Batch and continuous lipase-catalyzed interesterification of blends containing olive oil for trans-free margarines. Euro J Lipid Sci Technol 115(4):413–428. https://doi.org/10.1002/ejlt.201200418
19. Pande G, Akoh CC (2013) Enzymatic synthesis of trans-free structured margarine fat analogs with high stearate soybean oil and palm stearin and their characterization. LWT Food Sci Technol 50(1):232–239. https://doi.org/10.1016/j.lwt.2012.05.027
20. Ruan X, Zhu X, Xiong H, Wang S, Bai C, Zhao Q (2014) Characterisation of zero-trans margarine fats produced from camellia seed oil, palm stearin and coconut oil using enzymatic interesterification strategy. Int J Food Sci Technol 49(1):91–97. https://doi.org/10.1111/ijfs.12279
21. Pang M, Ge Y, Cao L, Cheng J, Jiang S (2019) Physicochemical properties, crystallization behavior and oxidative stabilities of enzymatic interesterified fats of beef tallow, palm stearin and camellia oil blends. J Oleo Sci 68(2):131–139. https://doi.org/10.5650/jos.ess18201
22. Kadivar S, De Clercq N, Mokbul M, Dewettinck K (2016) Influence of enzymatically produced sunflower oil based cocoa butter equivalents on the phase behavior of cocoa butter and quality of dark chocolate. LWT Food Sci Technol 66:48–55. https://doi.org/10.1016/j.lwt.2015.10.006
23. Haque Akanda J et al (2020) Hard fats improve the physicochemical and thermal properties of seed fats for applications in confectionery products. Food Rev Int 36(6):601–625. https://doi.org/10.1080/87559129.2019.1657443

24. Yamoneka J, Malumba P, Lognay G, Béra F, Blecker C, Danthine S (2018) Enzymatic inter-esterification of binary blends containing *Irvingia gabonensis* seed fat to produce cocoa butter substitute. Euro J Lipid Sci Technol 120(4). https://doi.org/10.1002/ejlt.201700423

25. Bahari A, Akoh CC (2018) Texture, rheology and fat bloom study of 'chocolates' made from cocoa butter equivalent synthesized from illipe butter and palm mid-fraction. LWT 97:349–354. https://doi.org/10.1016/j.lwt.2018.07.013

26. Bootello MA, Hartel RW, Garcés R, Martínez-Force E, Salas JJ (2012) Evaluation of high oleic-high stearic sunflower hard stearins for cocoa butter equivalent formulation. Food Chem 134(3):1409–1417. https://doi.org/10.1016/j.foodchem.2012.03.040

27. de Oliveira GM, Badan Ribeiro AP, dos Santos AO, Cardoso LP, Kieckbusch TG (2015) Hard fats as additives in palm oil and its relationships to crystallization process and polymorphism. LWT-Food Sci Technol 63(2):1163–1170. https://doi.org/10.1016/j.lwt.2015.04.036

28. Zhang Z, Song J, Lee WJ, Xie X, Wang Y (2020) Characterization of enzymatically interester-ified palm oil-based fats and its potential application as cocoa butter substitute. Food Chem 318:126518. https://doi.org/10.1016/j.foodchem.2020.126518

29. Ornla-ied P, Rungsang S, Tan CP, Lan D, Wang Y, Sonwai S (2022) Production of cocoa butter substitute via enzymatic interesterification of fully hydrogenated palm kernel oil, coconut oil and fully hydrogenated palm stearin blends. J Oleo Sci 71(3):ess21277. https://doi.org/10.5650/jos.ess21277

30. Cao H, Gerhold K, Mayers JR, Wiest MM, Watkins SM, Hotamisligil GS (2008) Identification of a lipokine, a lipid hormone linking adipose tissue to systemic metabolism. Cell 134(6):933–944. https://doi.org/10.1016/j.cell.2008.07.048

31. Moreira DKT, Santos PS, Gambero A, Macedo GA (2017) Evaluation of structured lipids with behenic acid in the prevention of obesity. Food Res Int 95:52–58. https://doi.org/10.1016/j.foodres.2017.03.005

32. Sivakanthan S, Jayasooriya AP, Madhujith T (2019) Optimization of the production of struc-tured lipid by enzymatic interesterification from coconut (Cocos nucifera) oil and sesame (Sesamum indicum) oil using response surface methodology. LWT 101:723–730. https://doi.org/10.1016/j.lwt.2018.11.085

33. Zam W (2020) Structured lipids: synthesis, health effects, and nutraceutical applications. In: Lipids and edible oils, Elsevier, pp 289–327. https://doi.org/10.1016/B978-0-12-817105-9.00008-2

34. Straarup EM, Høy C-E (2000) Structured lipids improve fat absorption in normal and malabsorbing rats. J Nutr 130(11):2802–2808. https://doi.org/10.1093/jn/130.11.2802

35. Chen J et al (2013) Structured lipid emulsion as nutritional therapy for the elderly patients with severe sepsis. Chin Med J (Engl) 126(12):2329–2332. https://doi.org/10.3760/cma.j.issn.0366-6999.20130758

36. Kanjilal S et al (2013) Hypocholesterolemic effects of low calorie structured lipids on rats and rabbits fed on normal and atherogenic diet. Food Chem 136(1):259–265. https://doi.org/10.1016/j.foodchem.2012.07.116

37. Moreira DKT et al (2020) Synthesis and characterization of structured lipid rich in behenic acid by enzymatic interesterification. Food Bioprod Process 122:303–310. https://doi.org/10.1016/j.fbp.2020.06.005

38. Moreira DKT, Ract JNR, Ribeiro APB, Macedo GA (2017) Production and characterization of structured lipids with antiobesity potential and as a source of essential fatty acids. Food Res Int 99:713–719. https://doi.org/10.1016/j.foodres.2017.06.034

39. da Silva RM et al (2020) Structured lipid containing behenic acid versus orlistat for weight loss: an experimental study in mice. PharmaNutrition 14:100213. https://doi.org/10.1016/j.phanu.2020.100213

40. de Gandra RL et al (2021) Production and characterization of nanoemulsion with low-calorie structured lipids and its potential to modulate biomarkers associated with obesity and comorbidities. Food Res Int 150:110782. https://doi.org/10.1016/j.foodres.2021.110782

41. Kim BH, Akoh CC (2006) Characteristics of structured lipid prepared by lipase-catalyzed acidolysis of roasted sesame oil and caprylic acid in a bench-scale continuous packed bed reactor. J Agric Food Chem 54(14):5132–5141. https://doi.org/10.1021/jf060607k

42. Choi J et al (2012) Lipase-catalysed production of triacylglycerols enriched in pinolenic acid at the *sn*-2 position from pine nut oil. J Sci Food Agric 92(4):870–876. https://doi.org/10.1002/jsfa.4662

43. Lee Y, Tang T, Lai O (2012) Health benefits, enzymatic production, and application of medium- and long-chain triacylglycerol (MLCT) in food industries: a review. J Food Sci 77(8). https://doi.org/10.1111/j.1750-3841.2012.02793.x

44. Casas-Godoy L, Marty A, Sandoval G, Ferreira-Dias S (2013) Optimization of medium chain length fatty acid incorporation into olive oil catalyzed by immobilized Lip2 from Yarrowia lipolytica. Biochem Eng J 77:20–27. https://doi.org/10.1016/j.bej.2013.05.001

45. Caballero E, Soto C, Olivares A, Altamirano C (2014) Potential use of avocado oil on structured lipids MLM-type production catalysed by commercial immobilised lipases. PLoS ONE 9(9):e107749. https://doi.org/10.1371/journal.pone.0107749

46. Costa CM et al (2018) Production of MLM type structured lipids from grapeseed oil catalyzed by non-commercial lipases. Eur J Lipid Sci Technol 120(1):1700320. https://doi.org/10.1002/ejlt.201700320

47. Sousa V, Campos V, Nunes P, Pires-Cabral P (2018) Incorporation of capric acid in pumpkin seed oil by *sn*-1,3 regioselective lipase-catalyzed acidolysis. OCL 25(3):A302. https://doi.org/10.1051/ocl/2018004

48. Suleiman JB, Nna VU, Zakaria Z, Othman ZA, Bakar ABA, Mohamed M (2020) Obesity-induced testicular oxidative stress, inflammation and apoptosis: protective and therapeutic effects of orlistat. Reprod Toxicol 95:113–122. https://doi.org/10.1016/j.reprotox.2020.05.009

49. Costin GM, Segal R (1999) Alimente functionale. Academica, Galati

50. Heydinger JA, Nakhasi DK (1996) Medium chain triacylglycerols. J Food Lipids 3(4):251–257. https://doi.org/10.1111/j.1745-4522.1996.tb00072.x

51. Albracht-Schulte K et al (2018) Omega-3 fatty acids in obesity and metabolic syndrome: a mechanistic update. J Nutr Biochem 58:1–16. https://doi.org/10.1016/j.jnutbio.2018.02.012

52. Çakmur H (2020) Introductory chapter: unbearable burden of the diseases—obesity. In: Obesity, IntechOpen. https://doi.org/10.5772/intechopen.85234

53. Ruiz-Rodriguez A, Reglero G, Ibañez E (2010) Recent trends in the advanced analysis of bioactive fatty acids. J Pharm Biomed Anal 51(2):305–326. https://doi.org/10.1016/j.jpba.2009.05.012

54. Kleiner L, Akoh CC (2018) Applications of structured lipids in selected food market segments and their evolving consumer demands. In: Lipid modification by enzymes and engineered microbes, Elsevier, pp 179–202. https://doi.org/10.1016/B978-0-12-813167-1.00009-8

55. Osório NM, Dubreucq E, da Fonseca MMR, Ferreira-Dias S (2009) Lipase/acyltransferase-catalysed interesterification of fat blends containing *n*-3 polyunsaturated fatty acids. Eur J Lipid Sci Technol 111(2):120–134. https://doi.org/10.1002/ejlt.200800109

56. Farfán M, Álvarez A, Gárate A, Bouchon P (2015) Comparison of chemical and enzymatic interesterification of fully hydrogenated soybean oil and walnut oil to produce a fat base with adequate nutritional and physical characteristics. Food Technol Biotechnol 53. https://doi.org/10.17113/ftb.53.03.15.3854

57. Akanbi TO, Barrow CJ (2015) Lipase-catalysed incorporation of EPA into emu oil: formation and characterisation of new structured lipids. J Funct Foods 19:801–809. https://doi.org/10.1016/j.jff.2014.11.010

58. Paula AV, Nunes GFM, Osório NM, Santos JC, de Castro HF, Ferreira-Dias S (2015) Continuous enzymatic interesterification of milkfat with soybean oil produces a highly spreadable product rich in polyunsaturated fatty acids. Eur J Lipid Sci Technol 117(5):608–619. https://doi.org/10.1002/ejlt.201400316

59. Lei Q, Ba S, Zhang H, Wei Y, Lee JY, Li T (2016) Enrichment of omega-3 fatty acids in cod liver oil via alternate solvent winterization and enzymatic interesterification. Food Chem 199:364–371. https://doi.org/10.1016/j.foodchem.2015.12.005

60. Chen Y, Liu K, Yang Z, Chang M, Wang X, Wang X (2023) Lipase-catalyzed two-step hydrolysis for concentration of acylglycerols rich in ω-3 polyunsaturated fatty acids. Food Chem 400:134115. https://doi.org/10.1016/j.foodchem.2022.134115

61. Lemaitre-Delaunay D, Pachiaudi C, Laville M, Pousin J, Armstrong M, Lagarde M (1999) Blood compartmental metabolism of docosahexaenoic acid (DHA) in humans after ingestion of a single dose of [(13)C]DHA in phosphatidylcholine. J Lipid Res 40(10):1867–1874

62. Guo Z, Vikbjerg AF, Xu X (2005) Enzymatic modification of phospholipids for functional applications and human nutrition. Biotechnol Adv 23(3):203–259. https://doi.org/10.1016/j.biotechadv.2005.02.001

63. Kim BH, Akoh CC (2015) Recent research trends on the enzymatic synthesis of structured lipids. J Food Sci 80(8). https://doi.org/10.1111/1750-3841.12953

64. Zhao T, No DS, Kim BH, Garcia HS, Kim Y, Kim I-H (2014) Immobilized phospholipase A1-catalyzed modification of phosphatidylcholine with n−3 polyunsaturated fatty acid. Food Chem 157:132–140. https://doi.org/10.1016/j.foodchem.2014.02.024

65. Kim JH, Yoon SH (2014) Effects of organic solvents on transesterification of phospholipids using phospholipase A2 and lipase. Food Sci Biotechnol 23(4):1207–1211. https://doi.org/10.1007/s10068-014-0165-6

66. Vikbjerg AF, Mu H, Xu X (2007) Synthesis of structured phospholipids by immobilized phospholipase A2 catalyzed acidolysis. J Biotechnol 128(3):545–554. https://doi.org/10.1016/j.jbiotec.2006.11.006

67. Taguchi H et al (2000) Double-blind controlled study on the effects of dietary diacylglycerol on postprandial serum and chylomicron triacylglycerol responses in healthy humans. J Am Coll Nutr 19(6):789–796. https://doi.org/10.1080/07315724.2000.10718079

68. Michalski MC et al (2013) Multiscale structures of lipids in foods as parameters affecting fatty acid bioavailability and lipid metabolism. Prog Lipid Res 52(4):354–373. https://doi.org/10.1016/j.plipres.2013.04.004

69. Panikashvili D et al (2006) The endocannabinoid 2-AG protects the blood–brain barrier after closed head injury and inhibits mRNA expression of proinflammatory cytokines. Neurobiol Dis 22(2):257–264. https://doi.org/10.1016/j.nbd.2005.11.004

70. Blankman JL, Simon GM, Cravatt BF (2007) A comprehensive profile of brain enzymes that hydrolyze the endocannabinoid 2-arachidonoylglycerol. Chem Biol 14(12):1347–1356. https://doi.org/10.1016/j.chembiol.2007.11.006

71. Eom T-K, Kong C-S, Byun H-G, Jung W-K, Kim S-K (2010) Lipase catalytic synthesis of diacylglycerol from tuna oil and its anti-obesity effect in C57BL/6J mice. Process Biochem 45(5):738–743. https://doi.org/10.1016/j.procbio.2010.01.012

72. Ferreira-Dias S, Correia AC, da Fonseca MMR (2003) Response surface modeling of glycerolysis catalyzed by Candida rugosa lipase immobilized in different polyurethane foams for the production of partial glycerides. J Mol Catal B Enzym 21(1–2):71–80. https://doi.org/10.1016/S1381-1177(02)00142-X

73. Liao H-F, Tsai W-C, Chang S-W, Shieh C-J (2003) Application of solvent engineering to optimize lipase-catalyzed 1,3-diglyacylcerols by mixture response surface methodology. Biotechnol Lett 25(21):1857–1861. https://doi.org/10.1023/A:1026237829284

74. Mangas-Sánchez J, Serrano-Arnaldos M, Adlercreutz P (2015) Effective and highly selective lipase-mediated synthesis of 2-monoolein and 1,2-diolein in a two-phase system. J Mol Catal B Enzym 112:9–14. https://doi.org/10.1016/j.molcatb.2014.11.014

75. Wang X, Li M, Wang T, Jin Q, Wang X (2014) An improved method for the synthesis of 2-arachidonoylglycerol. Process Biochem 49(9):1415–1421. https://doi.org/10.1016/j.procbio.2014.05.021

76. Dhara R, Singhal RS (2014) Process optimization of enzyme catalyzed production of dietary diacylglycerol (DAG) using TLIM as biocatalyst. J Oleo Sci 63(2):169–176. https://doi.org/10.5650/jos.ess13159

77. Zhou J, Lee YY, Mao Y, Wang Y, Zhang Z (2002) Future of structured lipids: enzymatic synthesis and their new applications in food systems. Foods 11(16):2400. https://doi.org/10.3390/foods11162400

78. Lo SK, Tan CP, Long K, Yusoff SA, Lai OM (2008) Diacylglycerol oil—properties, processes and products: a review. Food Bioproc Tech 1(3):223–233. https://doi.org/10.1007/s11947-007-0049-3

79. Diao X, Guan H, Kong B, Zhao X (2017) Preparation of diacylglycerol from lard by enzymatic glycerolysis and its compositional characteristics. Korean J Food Sci Anim Resour 37(6):813
80. Li D et al (2021) Simultaneous preparation of edible quality medium and high purity diacyl-glycerol by a novel combined approach. LWT 150:111949. https://doi.org/10.1016/j.lwt.2021. 111949
81. Zhao X, Sun Q, Qin Z, Liu Q, Kong B (2018) Ultrasonic pretreatment promotes diacylglyc-erol production from lard by lipase-catalysed glycerolysis and its physicochemical properties. Ultrason Sonochem 48:11–18. https://doi.org/10.1016/j.ultsonch.2018.05.005
82. Lee Y-Y et al (2020) Production, safety, health effects and applications of diacylglycerol func-tional oil in food systems: a review. Crit Rev Food Sci Nutr 60(15):2509–2525. https://doi.org/ 10.1080/10408398.2019.1650001
83. Bechtel DH (2014) Article series: safety of esterified propoxylated glycerol (EPG), a nonab-sorbable fat replacer. Regul Toxicol Pharmacol 70:S91–S94. https://doi.org/10.1016/j.yrtph. 2014.11.010
84. Yanai H, Tomono Y, Ito K, Furutani N, Yoshida H, Tada N (2007) Diacylglycerol oil for the metabolic syndrome. Nutr J 6(1):43. https://doi.org/10.1186/1475-2891-6-43
85. González A, Martínez ML, Paredes AJ, León AE, Ribotta PD (2016) Study of the preparation process and variation of wall components in chia (Salvia hispanica L.) oil microencapsulation. Powder Technol 301:868–875. https://doi.org/10.1016/j.powtec.2016.07.026
86. Sandhya K, Leena MM, Moses JA, Anandharamakrishnan C (2023) Edible oil to powder technologies: concepts and advances. Food Biosci 53:102567. https://doi.org/10.1016/j.fbio. 2023.102567
87. Chen X-W, Yang X-Q (2019) Characterization of orange oil powders and oleogels fabricated from emulsion templates stabilized solely by a natural triterpene saponin. J Agric Food Chem 67(9):2637–2646. https://doi.org/10.1021/acs.jafc.8b04588
88. Hosseini F, Miri MA, Najafi M, Soleimanifard S, Aran M (2021) Encapsulation of rosemary essential oil in zein by electrospinning technique. J Food Sci 86(9):4070–4086. https://doi.org/ 10.1111/1750-3841.15876
89. Elik A, Koçak Yanık D, Göğüş F (2021) A comparative study of encapsulation of carotenoid enriched-flaxseed oil and flaxseed oil by spray freeze-drying and spray drying techniques. LWT 143:111153. https://doi.org/10.1016/j.lwt.2021.111153
90. Plati F, Papi R, Paraskevopoulou A (2021) Characterization of oregano essential oil (Origanum vulgare L. subsp. hirtum) particles produced by the novel nano spray drying technique. Foods 10(12):2923. https://doi.org/10.3390/foods10122923
91. Thuong Nhan NP et al (2020) Microencapsulation of lemongrass (Cymbopogon citratus) essen-tial oil via spray drying: effects of feed emulsion parameters. Processes 8(1):40. https://doi. org/10.3390/pr8010040
92. Reineccius G, Patil S, Anantharamkrishnan V (2022) Encapsulation of orange oil using fluidized bed granulation. Molecules 27(6):1854. https://doi.org/10.3390/molecules27061854
93. Pereira de Oliveira J et al (2022) Tailoring the physicochemical properties of freeze-dried buriti oil microparticles by combining inulin and gum Arabic as encapsulation agents. LWT 161:113372. https://doi.org/10.1016/j.lwt.2022.113372
94. Beacom E, Bogue J, Repar L (2021) Market-oriented development of plant-based food and beverage products: a usage segmentation approach. J Food Product Market 27(4):204–222. https://doi.org/10.1080/10454446.2021.1955799
95. Rashid R, Wani SM, Manzoor S, Masoodi FA, Dar MM (2022) Improving oxidative stability of edible oils with nanoencapsulated orange peel extract powder during accelerated shelf life storage. Food Biosci 49:101917. https://doi.org/10.1016/j.fbio.2022.101917
96. Srivastava S, Mishra HN (2021) Development of microencapsulated vegetable oil powder based cookies and study of its physicochemical properties and storage stability. LWT 152:112364. https://doi.org/10.1016/j.lwt.2021.112364
97. Sharma S, Cheng S-F, Bhattacharya B, Chakkaravarthi S (2019) Efficacy of free and encapsu-lated natural antioxidants in oxidative stability of edible oil: special emphasis on nanoemulsion-based encapsulation. Trends Food Sci Technol 91:305–318. https://doi.org/10.1016/j.tifs.2019. 07.030

98. Weisany W et al (2022) Targeted delivery and controlled released of essential oils using nanoencapsulation: a review. Adv Colloid Interface Sci 303:102655. https://doi.org/10.1016/j.cis.2022.102655

Chapter 5
Structured Lipids: Importance in Metabolism and Health

5.1 Introduction

Lipids are considered vital nutrients for normal physiological activities and the maintenance of human health, providing essential fatty acids (FAs) [1]. However, it is well known that excessive consumption of saturated FAs is associated with an increased risk of obesity, which can consequently exacerbate and/or stimulate metabolic disorders related to metabolic syndrome and existing or pre-existing conditions corresponding to non-communicable chronic diseases (NCDs) [2–4].

Based on this and with the understanding of the negative impact of trans-fatty acids (TFAs) on health, legislation and guidelines from health and food agencies have emerged. These aim to limit the consumption of major sources of saturated fats to less than 10% of total caloric intake, along with the prohibition of the use of partially hydrogenated fats by the food industry [5–7]. Thus, recommendations emerged to replace these fats with oils and fats rich in monounsaturated and polyunsaturated FAs (MUFAs and PUFAs) and long-chain FAs, which are characteristic of the Mediterranean diet [8].

Structured lipids (SLs) are capable of precisely modifying fats and oils, either by reorganizing FAs to those with desired physical properties or by increasing the content of PUFAs or short-, medium-, and long-chain FAs. This provides unique functional properties to the lipids, such as easy absorption, low calorie content, and low saturated fat content, for example [1]. To achieve this, the catalyst used must be specific to direct these modifications, such as in lipase-catalyzed interesterification synthesis; these characteristics are not achieved by chemical methods due to their randomness [9].

It is noteworthy that SLs can have applications both in the food industry and in clinical nutrition, where interesterification synthesis is used to produce specialized foods for infants and patients suffering from lipid malabsorption disorders [10]. These applications also include their use in enteral and parenteral nutrition [11].

V. Alves et al., *Chemical and Enzymatic Interesterification for Food Lipid Production*,
Chemistry of Foods, https://doi.org/10.1007/978-3-031-67405-1_5

Additionally, interesterification has been used to design fats with low absorption for weight control, known as low-calorie SLs [12].

However, information on clinical trials in humans conducted with SLs and their relationship to health is still limited. In this context, this chapter aims to provide a better understanding of the effects of SLs on metabolism and human health.

5.2   The Role of Lipids in Human Nutrition

A balanced lipid diet with the consumption of oils and vegetable fats in adequate quantities, along with healthy eating, is essential for maintaining the human body. This is due to the need to consume certain FAs that the body does not produce, known as essential FAs, in sufficient quantities to meet its needs [13]. Omega-6 linoleic acid (C18:2 omega-6) and omega-3 linolenic acid (C18:3 omega-3) are considered essential FAs because they are not synthesized by the human body and are absolutely necessary for human health, meaning they must be included in the diet [14].

Linoleic acid in the human body can act as a precursor to the omega-3 family, meaning it can be converted into arachidonic acid (C20:4 omega-6) and alpha-linolenic acid into longer-chain omega-3 FAs through chain elongation and desaturation bioreactions, forming eicosapentaenoic acid (EPA; C20:5 omega-3) and docosahexaenoic acid (DHA; C22:6 omega-3) [13]. It is worth noting that EPA and DHA are primarily found in marine oils, providing significant health benefits. These FAs are important in the prevention and treatment of cardiovascular diseases, regulation of blood pressure, and inflammatory processes [15]. DHA has proven to be crucial in the infantile brain and nervous system development and their normal function in adulthood [16, 17].

Among the long-chain FAs with important nutritional, physiological, and/or physical properties are conjugated linoleic acid (CLA) and gamma-linolenic acid (GLA) [14]. CLA is found in the milk and meat of ruminants, and its consumption is associated with anti-carcinogenic, antioxidant, anti-atherosclerotic, anti-diabetic, anti-obesity, and immune-enhancing effects [18]. On the other hand, GLA is found in oilseeds from plants such as *Borago officinalis* L. and *Oenothera biennis* L., and it is associated with anti-tumor effects. It is a precursor to prostaglandin 1 and 15-hydroxyeicosatrienoic acid, which possess anti-inflammatory and antithrombotic properties [19].

The Western diet is considered insufficient in omega-3 FAs with an excessive intake of omega-6, considering genetic patterns and low physical activity frequency (sedentary lifestyle), which promotes the incidence of NCDs [20, 21]. Conversely, replacing a diet rich in saturated fats with MUFAs presents health benefits such as potential anti-atherogenic and antithrombotic effects. It increases the HDL/LDL (high-density lipoprotein/low-density lipoprotein) cholesterol ratio, reduces oxidized LDL and total cholesterol, decreases platelet aggregation, and improves insulin sensitivity [22, 23].

5.3 Lipid Absorption and Metabolism

The composition of triacylglycerols (TAGs) of FAs, their position in the acylglycerol structure, and their physical and biochemical properties, as well as chain length, result in different physiological properties demonstrated in terms of absorption in the gastrointestinal tract and metabolic processes [24]. FAs located at positions *sn*-1 and *sn*-3 may have different metabolic fates than FAs at position *sn*-2 [12].

The first step in fat digestion occurs in the stomach and is catalyzed by lingual or gastric lipase, with the main products of gastric phase digestion being diacylglycerols (DAGs) and free fatty acids (FFAs), which facilitate the intestinal digestion phase [25]. In the duodenal region, hydrolysis is catalyzed by pancreatic lipase, which is regioselective at positions *sn*-1,3 with a higher affinity for *sn*-1 than *sn*-3 positions of TAGs, generating FFAs and sn-2 monoacylglycerols (sn-2 MAG), which form micelles with bile salts [26]. When these micelles, also containing phospholipids, approach the apical side of intestinal epithelial cells, they release their contents and allow for the absorption of nonpolar lipids into the membrane of microvilli. The absorbed lipids are re-esterified, forming TAG again in the smooth endoplasmic reticulum [27]. It is worth noting that the hydrolysis reaction rate performed by pancreatic lipase depends on the chain length and degree of unsaturation of the FA present at positions *sn*-1 and *sn*-3 of TAG [24].

The absorption of 2-MAGs occurs through passive diffusion into enterocytes [28], where they are first recycled for the synthesis of new TAGs and reassembled into lymphatic chylomicrons at the level of the endoplasmic reticulum, maintaining their native stereo-specificity [11, 29, 30]. Lipoprotein lipase (LPL), responsible for the highly efficient hydrolysis of TAG-chylomicrons, is also specific to the *sn*-1 and *sn*-3 positions, like pancreatic lipase, so the obtained 2-MAG is isomerized into 1(3)-MAG from which FA is efficiently released [31]. Absorbed MAGs can also serve as a primary structure for the synthesis of phospholipids in the intestine or liver in environments with excess FFAs [11].

Some MUFAs and PUFAs can be esterified at the *sn*-2 position of TAG, which promotes their complete absorption at the *sn*-2 position, like sn-2 MAG, which is essential for humans and is absorbed by the mucous membrane in the intestinal epithelium in the form of micelles with bile salts, thereby avoiding its deficiency in the body [32]. Regarding the absorption of FAs, short-chain ones are absorbed more rapidly in the stomach than other FAs with longer chains due to their high volatility, water solubility, and low molecular weight, making them also suitable for obesity control due to their low caloric value [33].

Short-chain and medium-chain FAs, originating from *sn*-1 and *sn*-3 positions, can be absorbed in the stomach after hydrolysis by gastric lipase [34], or they can be solubilized in the aqueous phase of intestinal content, where they are weakly bound to albumin and transported by the portal system to the liver for oxidation [12]. Here, they are almost entirely taken up by the liver, with the remainder passing into the bloodstream, available to peripheral tissues, as they are not significantly incorporated into lipoproteins (chylomicrons and VLDL—very low-density lipoproteins), allowing for their direct absorption into the bloodstream [13]. Since they are rapidly oxidized in the liver and utilized, they undergo minimal re-esterification in TAG, allowing for their anti-obesogenic effect as they do not accumulate in adipose tissue [35]. Thus, medium-chain FAs have been used as a quick energy source in some metabolic syndromes, such as pancreatic enzyme deficiency (cystic fibrosis) [36]. However, it is worth noting that medium-chain FAs should be part of a balanced diet, as they are saturated FAs. On the other hand, PUFAs located at the *sn*-1 and *sn*-3 positions limit the hydrolytic activity of pancreatic lipase due to their steric hindrance and thus influence the kinetics of FA release [37].

Regarding long-chain FAs, these exhibits high hydrophobicity resulting from their long hydrocarbon chains, which hinders their direct absorption and/or transportation [13]. Long-chain FAs exit the intestine in the form of TAGs via the lymphatic system after incorporation into chylomicrons, which are formed by TAGs, phospholipids, cholesterol, and apoproteins and are eventually secreted into the bloodstream [36]. Additionally, a fraction of these chylomicrons undergoes intravascular hydrolysis, releasing the majority of these long-chain FAs to extrahepatic tissues, while the remaining fraction is transported to the liver, where they arrive in the form of FAs bound to albumin or in the form of TAGs [13]. The longer the carbon chain of the FA, the more it is found in the lymph (TAGs associated with chylomicrons) and less in the portal blood (FAs bound to albumin) [38].

5.4 Metabolism of Interesterified Lipids

The structuring of FAs in TAG via interesterification determines the bioavailability of FAs. Specifically, FAs at the *sn*-1 and *sn*-3 positions of TAG are preferentially hydrolyzed by pancreatic and lipoprotein lipases. This is unlike saturated FAs with high melting points (such as stearic and palmitic acids), which are poorly absorbed as FFAs compared to when they are present as 2-MAGs [39].

In vivo studies have shown that the absorption of fat from vegetable oils containing saturated FAs at the *sn*-1 and *sn*-3 positions occurs at a lower rate compared to saturated FAs at the *sn*-2 position [40, 41]. This could be explained by the fact that long-chain saturated FAs, such as palmitic acid, have low absorption coefficients due to their melting points above body temperature. Additionally, there is a difference in polarity and capacity in the formation of micelles with bile acids and in the formation of insoluble calcium soaps. Long-chain saturated FAs released from *sn*-1,3 positions in the intestinal lumen tend to bind to Ca and Mg ions to form insoluble salts due to the alkaline intestinal environment [42]. However, when medium-chain FAs derived from interesterification synthesis are found at the *sn*-2 position, a higher proportion is transported as lymphatic chylomicrons because they are absorbed as MAG, which is re-esterified into TAGs by enterocytes [24]. Therefore, interesterification synthesis of fats and vegetable oils can increase the proportion of saturated FAs at the *sn*-2 position, improving their absorption/digestibility [43].

This condition is relevant for infant nutrition in terms of fat and Ca absorption, as the absorption of fat from human milk has a higher rate than that of infant formula, partly due to the preferential presence of palmitic acid at the sn-2 position of the glycerol TAG structure [44]. In fact, in infants, once absorbed, palmitic acid is esterified into TAG and secreted into the plasma. Therefore, the specific sn-2 position of palmitic acid in human milk may influence the formation of chylomicrons after digestion and also the metabolism of cholesterol esters and long-chain PUFAs [45]. In contrast to human milk, in conventional infant formula, palmitic acid is mainly located at the sn-1 and sn-3 positions and is absorbed as non-esterified FAs. To increase fat absorption similarly to human breast milk, infant formula products (Betapol™) contain TAGs with up to 60% or more of palmitic acid at the sn-2 position [36, 46, 47].

Refrences [48–50] evaluated medium- and long-chain FAs in vivo trials and observed that these FAs were transported more rapidly to the lymph from the sn-2 position than from the sn-1,3 positions, providing high energy to peripheral tissues that could readily be utilized, which would otherwise be predominantly transported to the liver through the portal circulation. Refrences [51, 52] also supported in their in vivo studies and clinical trials that interesterified lipids are capable of providing medium-chain FAs to peripheral tissues in an energetic manner.

Teo et al. [50] discuss that SLs can improve the absorption of FAs into the lymph of rats after ischemia–reperfusion injury. References [53, 54], argue that the use of SLs in rat malabsorption models and canine models increased the absorption of fat-soluble vitamins and the bioavailability of lipophilic drugs. References [51, 52], suggest that in patients with high physical stress, such as burns or surgical trauma, interesterified fats may be beneficial in reducing nitrogen loss and organ damage.

It's worth noting that most of the literature available on clinical trials lacks information on trials conducted with interesterified lipids in humans. Shagholian et al. [55] discuss that SLs composed of equal amounts of CLA and conjugated linoleic acid (CLnA) showed potential in reducing liver lipid and TAG levels. Despite [56] observing favorable effects of CLA on cardiometabolic health in an observational study in humans, there remains a need for more clinical evidence from randomized clinical trials, and those available have reported unfavorable outcomes regarding lipid peroxidation and inflammation using CLA [57–59].

Mensink et al. [24] studied the long-term metabolic effects of interesterified lipids, and most of the data suggested only limited differences in the long-term effects of long-chain saturated FAs from native TAGs compared to interesterified lipids in fasting serum lipoproteins, highlighting the need for more long-term clinical trials to determine safety issues related to the use of structured lipids in humans.

The metabolism and absorption of TAGs are illustrated in Fig. 5.1.

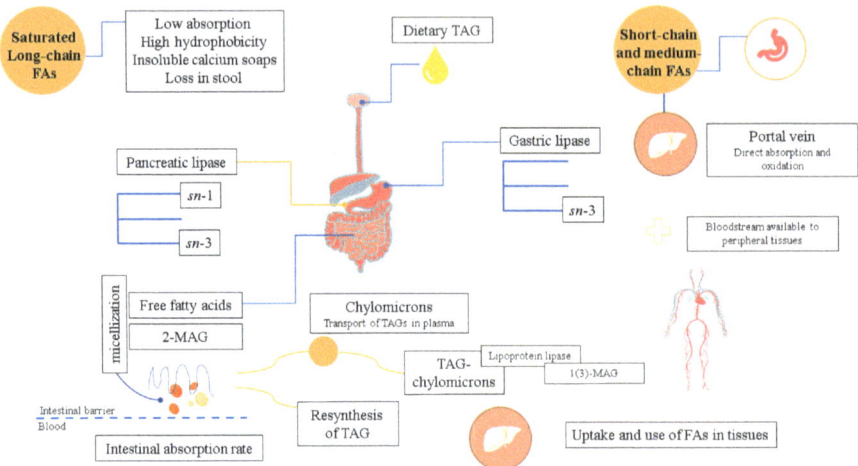

Fig. 5.1 Digestion and absorption of dietary TAGs. Dietary TAGs are digested to obtain FFAs and 2-MAGs in the intestinal lumen, which intestinal epithelial cells absorb. Here, 2-MAGs rapidly resynthesize TAGs which are included in chylomicron by lipoprotein lipase and transported in the intestinal lymph. FFAs, via the glycerol-3-phosphate pathway, are transported by the portal system to the liver for oxidation Figure adapted from Ref. [37]

5.5 Structured Lipids for Clinical Applications

SLs developed from the rearrangement of FAs in TAGs have provided lipids with specific nutritional value and health effects. Thus, these lipids can consist of high proportions of highly unsaturated FAs such as EPA (C20:5n-3), DHA (C22:6n-3), linolenic acid (C18:3n-3), and gamma-linolenic acids (GLA, C18:3n-6), considered essential FAs for nutritional and pharmaceutical applications [10]. It is worth noting that omega-3 FAs are associated with improvements in cardiovascular and inflammatory diseases and optimal brain functionality [60]. However, it is important to note that an increase in the ratio of n-6/n-3 FAs is linked to an increased risk of cardiovascular diseases, necessitating an increase in omega-3 intake to reduce this ratio [61]. SLs derived from lipid modification emerge as a resource to reduce this omega-6/omega-3 ratio in the human diet [62].

Sivakanthan and Madhujith [10] discuss in a review that SLs containing medium- and long-chain FAs provide nutritional and therapeutic benefits. TAGs containing PUFAs in the *sn*-2 position and medium-chain FAs in the *sn*-1 and *sn*-3 positions are rapidly hydrolyzed by pancreatic lipase. Therefore, these SLs may be desirable as "nutraceuticals," as caproic, caprylic, and capric acids are not incorporated into human body lipoprotein groups and can be used to produce lipids for clinical purposes, such as supplementation in infant formulas or as a dietary supplement for adults. They can be easily oxidized in the liver of premature babies and patients with absorption problems, thus serving as an important source of energy. Additionally, [63] emphasize the benefit of using fish oil with medium-chain FAs in SLs to maintain substantial levels of DHA and EPA through enzymatic interesterification.

References

1. Zhou J, Lee Y-Y, Mao Y, Wang Y, Zhang Z (2022) Future of structured lipids: enzymatic synthesis and their new applications in food systems. Foods 11(16):2400. https://doi.org/10.3390/foods11162400
2. Albracht-Schulte K et al (2018) Omega-3 fatty acids in obesity and metabolic syndrome: a mechanistic update. J Nutr Biochem 58:1–16. https://doi.org/10.1016/j.jnutbio.2018.02.012
3. Çakmur H (2020) Introductory chapter: unbearable burden of the diseases—obesity. In: Obesity, IntechOpen. https://doi.org/10.5772/intechopen.85234
4. Hooper L, Martin N, Jimoh OF, Kirk C, Foster E, Abdelhamid AS (2020) Reduction in saturated fat intake for cardiovascular disease. Cochrane Database Syst Rev 8:2020. https://doi.org/10.1002/14651858.CD011737.pub3
5. David ML, Guivant JS (2012) A gordura trans: entre as controvérsias científicas e as estratégias da indústria alimentar. Política & Sociedad 11(20). https://doi.org/10.5007/2175-7984.2012v11n20p49
6. da S. Opas OPA Plano de Ação para Eliminar Os Ácidos Graxos Trans De Produção Industrial 2020–2025. https://www.paho.org/en/documents/plan-action-elimination-industrially-produced-trans-fatty-acids-2020-2025
7. Food and Drug Administration. Final determination regarding partially hydrogenated oils (Removing Trans Fat). https://www.fda.gov/food/food-additives-petitions/final-determination-regarding-partially-hydrogenated-oils-removing-trans-fat.
8. Basak S, Banerjee A, Pathak S, Duttaroy AK (2022) Dietary fats and the gut microbiota: their impacts on lipid-induced metabolic syndrome. J Funct Foods 91:105026. https://doi.org/10.1016/j.jff.2022.105026
9. Kadhum AAH, Shamma MN (2017) Edible lipids modification processes: a review. Crit Rev Food Sci Nutr 57(1):48–58. https://doi.org/10.1080/10408398.2013.848834
10. Sivakanthan S, Madhujith T (2020) Current trends in applications of enzymatic interesterification of fats and oils: a review. LWT 132:109880. https://doi.org/10.1016/j.lwt.2020.109880
11. Karupaiah T, Sundram K (2007) Effects of stereospecific positioning of fatty acids in triacylglycerol structures in native and randomized fats: a review of their nutritional implications. Nutr Metab (Lond) 4(1):16. https://doi.org/10.1186/1743-7075-4-16
12. Alfieri A et al (2017) Effects of plant oil interesterified triacylglycerols on lipemia and human health. Int J Mol Sci 19(1):104. https://doi.org/10.3390/ijms19010104
13. Ferreira-Dias S, Osório N, Tecelão C (2022) Bioprocess technologies for production of structured lipids as nutraceuticals. In: Current developments in biotechnology and bioengineering, Elsevier, pp 209–237. https://doi.org/10.1016/B978-0-12-823506-5.00007-2
14. Ferreira-Dias S (2010) Enzymatic production of functional fats. In: Comprehensive food fermentation biotechnology. India, Asiatech Publishers Inc 2, pp 608–641
15. Howlett J (2008) Functional foods: from science to health and claims
16. Valenzuela AB (2009) Docosahexaenoic acid (DHA), an essential fatty acid for the proper functioning of neuronal cells: their role in mood disorders. Grasas y Aceites 60(2):203–212. https://doi.org/10.3989/gya.085208
17. Helland IB, Smith L, Saarem K, Saugstad OD, Drevon CA (2003) Maternal supplementation with very-long-chain n-3 fatty acids during pregnancy and lactation augments children's iq at 4 years of age. Pediatrics 111(1):e39–e44. https://doi.org/10.1542/peds.111.1.e39

18. Goli SAH, Kadivar M, Keramat J, Fazilati M (2008) Conjugated linoleic acid (CLA) production and lipase-catalyzed interesterification of purified CLA with canola oil. Eur J Lipid Sci Technol 110(5):400–404. https://doi.org/10.1002/ejlt.200700267
19. Lumor SE, Akoh CC (2005) Incorporation of γ-linolenic and linoleic acids into a palm kernel oil/palm olein blend. Eur J Lipid Sci Technol 107(7–8):447–454. https://doi.org/10.1002/ejlt.200501157
20. DiNicolantonio JJ, O'Keefe JH (2018) Importance of maintaining a low omega–6/omega–3 ratio for reducing inflammation. Open Heart 5(2):e000946. https://doi.org/10.1136/openhrt-2018-000946
21. Simopoulos A (2016) An increase in the omega-6/omega-3 fatty acid ratio increases the risk for obesity. Nutrients 8(3):128. https://doi.org/10.3390/nu8030128
22. Rajaram S, Haddad EH, Mejia A, Sabaté J (2009) Walnuts and fatty fish influence different serum lipid fractions in normal to mildly hyperlipidemic individuals: a randomized controlled study. Am J Clin Nutr 89(5):1657S-1663S. https://doi.org/10.3945/ajcn.2009.26736S
23. Palomer X, González-Clemente JM, Blanco-Vaca F, Mauricio D (2008) Role of vitamin D in the pathogenesis of type 2 diabetes mellitus. Diabetes Obes Metab 10(3):185–197. https://doi.org/10.1111/j.1463-1326.2007.00710.x
24. Mensink RP, Sanders TA, Baer DJ, Hayes K, Howles PN, Marangoni A (2016) The increasing use of interesterified lipids in the food supply and their effects on health parameters. Adv Nutr 7(4):719–729. https://doi.org/10.3945/an.115.009662
25. Versleijen M, Roelofs H, Preijers F, Roos D, Wanten G (2005) Parenteral lipids modulate leukocyte phenotypes in whole blood, depending on their fatty acid composition. Clin Nutr 24(5):822–829. https://doi.org/10.1016/j.clnu.2005.05.003
26. Ajuwon KM, Spurlock ME (2005) Palmitate activates the NF-κB transcription factor and induces IL-6 and TNFα expression in 3T3-L1 adipocytes. J Nutr 135(8):1841–1846. https://doi.org/10.1093/jn/135.8.1841
27. Gower RM et al (2011) CD11c/CD18 expression is upregulated on blood monocytes during hypertriglyceridemia and enhances adhesion to vascular cell adhesion molecule-1. Arterioscler Thromb Vasc Biol 31(1):160–166. https://doi.org/10.1161/ATVBAHA.110.215434
28. Schulthess G et al (1994) Absorption of monoacylglycerols by small intestinal brush border membrane. Biochemistry 33(15):4500–4508. https://doi.org/10.1021/bi00181a009
29. Berry SEE (2009) Triacylglycerol structure and interesterification of palmitic and stearic acid-rich fats: an overview and implications for cardiovascular disease. Nutr Res Rev 22(1):3–17. https://doi.org/10.1017/S0954422409369267
30. Yang LY, Kuksis A (1991) Apparent convergence (at 2-monoacylglycerol level) of phosphatidic acid and 2-monoacylglycerol pathways of synthesis of chylomicron triacylglycerols. J Lipid Res 32(7):1173–1186
31. Emken EA, Adlof RO, Duval SM, Shane JM, Walker PM, Becker C (2004) Effect of triacylglycerol structure on absorption and metabolism of isotope-labeled palmitic and linoleic acids by humans. Lipids 39(1). https://doi.org/10.1007/s11745-004-1194-6
32. Tholstrup T, Miller GJ, Bysted A, Sandström B (2003) Effect of individual dietary fatty acids on postprandial activation of blood coagulation factor VII and fibrinolysis in healthy young men. Am J Clin Nutr 77(5):1125–1132. https://doi.org/10.1093/ajcn/77.5.1125
33. Borkman M, Storlien LH, Pan DA, Jenkins AB, Chisholm DJ, Campbell LV (1993) The relation between insulin sensitivity and the fatty-acid composition of skeletal-muscle phospholipids. N Engl J Med 328(4):238–244. https://doi.org/10.1056/NEJM199301283280404
34. Risérus U, Willett WC, Hu FB (2009) Dietary fats and prevention of type 2 diabetes. Prog Lipid Res 48(1):44–51. https://doi.org/10.1016/j.plipres.2008.10.002
35. Berry SEE, Woodward R, Yeoh C, Miller GJ, Sanders TAB (2007) Effect of interesterification of palmitic acid-rich triacylglycerol on postprandial lipid and factor vii response. Lipids 42(4):315–323. https://doi.org/10.1007/s11745-007-3024-x
36. Sanders TA, Oakley FR, Cooper JA, Miller GJ (2001) Influence of a stearic acid–rich structured triacylglycerol on postprandial lipemia, factor VII concentrations, and fibrinolytic activity in healthy subjects. Am J Clin Nutr 73(4):715–721. https://doi.org/10.1093/ajcn/73.4.715

37. Michalski MC et al (2013) Multiscale structures of lipids in foods as parameters affecting fatty acid bioavailability and lipid metabolism. Prog Lipid Res 52(4):354–373. https://doi.org/10.1016/j.plipres.2013.04.004

38. Sanders TA, Filippou A, Berry SE, Baumgartner S, Mensink RP (2011) Palmitic acid in the sn-2 position of triacylglycerols acutely influences postprandial lipid metabolism. Am J Clin Nutr 94(6):1433–1441. https://doi.org/10.3945/ajcn.111.017459

39. Farfán M, Villalón MJ, Ortíz ME, Nieto S, Bouchon P (2013) The effect of interesterification on the bioavailability of fatty acids in structured lipids. Food Chem 139(1–4):571–577. https://doi.org/10.1016/j.foodchem.2013.01.024

40. Nagata J, Kasai M, Watanabe S, Ikeda I, Saito M (2003) Effects of highly purified structured lipids containing medium-chain fatty acids and linoleic acid on lipid profiles in rats. Biosci Biotechnol Biochem 67(9):1937–1943. https://doi.org/10.1271/bbb.67.1937

41. Porsgaard T, Høy C-E (2000) Lymphatic transport in rats of several dietary fats differing in fatty acid profile and triacylglycerol structure. J Nutr 130(6):1619–1624. https://doi.org/10.1093/jn/130.6.1619

42. Wang T, Wang X, Wang X (2016) Effects of lipid structure changed by interesterification on melting property and lipemia. Lipids 51(10):1115–1126. https://doi.org/10.1007/s11745-016-4184-3

43. Mills CE, Hall WL, Berry SEE (2017) What are interesterified fats and should we be worried about them in our diet? Nutr Bull 42(2):153–158. https://doi.org/10.1111/nbu.12264

44. Kallio H, Nylund M, Boström P, Yang B (2017) Triacylglycerol regioisomers in human milk resolved with an algorithmic novel electrospray ionization tandem mass spectrometry method. Food Chem 233:351–360. https://doi.org/10.1016/j.foodchem.2017.04.122

45. Ramírez M, Amate L, Gil A (2001) Absorption and distribution of dietary fatty acids from different sources. Early Hum Dev 65:S95–S101. https://doi.org/10.1016/S0378-3782(01)00211-0

46. Lucas A, Quinlan P, Abrams S, Ryan S, Meah S, Lucas PJ (1997) Randomised controlled trial of a synthetic triglyceride milk formula for preterm infants. Arch Dis Child Fetal Neonatal Ed 77(3):F178–F184. https://doi.org/10.1136/fn.77.3.F178

47. Miles EA, Calder PC (2017) The influence of the position of palmitate in infant formula triacylglycerols on health outcomes. Nutr Res 44:1–8. https://doi.org/10.1016/j.nutres.2017.05.009

48. Ikeda I, Tomari Y, Sugano M, Watanabe S, Nagata J (1991) Lymphatic absorption of structured glycerolipids containing medium-chain fatty acids and linoleic acid, and their effect on cholesterol absorption in rats. Lipids 26(5):369–373. https://doi.org/10.1007/BF02537201

49. Couëdelo L et al (2012) The fraction of α-linolenic acid present in the sn-2 position of structured triacylglycerols decreases in lymph chylomicrons and plasma triacylglycerols during the course of lipid absorption in rats3. J Nutr 142(1):70–75. https://doi.org/10.3945/jn.111.146290

50. Tso P, Lee T, Demichele SJ (1999) Lymphatic absorption of structured triglycerides vs. physical mix in a rat model of fat malabsorption. Am J Physiol Gastrointestin Liver Physiol 277(2):G333–G340. https://doi.org/10.1152/ajpgi.1999.277.2.G333

51. Teo TC, Demichele TJ, Selleck KM, Babayan VK, Blackburn GL, Bistrian BR (1989) Administration of structured lipid composed of MCT and fish oil reduces net protein catabolism in enterally fed burned rats. Ann Surg 210(1):100–107. https://doi.org/10.1097/00000658-198907000-00015

52. Kenler AS et al (1996) Early enteral feeding in postsurgical cancer patients. Ann Surg 223(3):316–333. https://doi.org/10.1097/00000658-199603000-00013

53. Tso P, Lee T, DeMichele SJ (2001) Randomized structured triglycerides increase lymphatic absorption of tocopherol and retinol compared with the equivalent physical mixture in a rat model of fat malabsorption. J Nutr 131(8):2157–2163. https://doi.org/10.1093/jn/131.8.2157

54. Holm R (2003) Examination of oral absorption and lymphatic transport of halofantrine in a triple-cannulated canine model after administration in self-microemulsifying drug delivery systems (SMEDDS) containing structured triglycerides. Eur J Pharm Sci 20(1):91–97. https://doi.org/10.1016/S0928-0987(03)00174-X

55. Shagholian M, Goli SAH, Shirvani A, Agha-Ghazvini MR, Asgary S (2019) Liver and serum lipids in Wistar rats fed a novel structured lipid containing conjugated linoleic acid and conjugated linolenic acid. Grasas y Aceites 70(2):307. https://doi.org/10.3989/gya.0582181
56. Smit LA, Baylin A, Campos H (2010) Conjugated linoleic acid in adipose tissue and risk of myocardial infarction. Am J Clin Nutr 92(1):34–40. https://doi.org/10.3945/ajcn.2010.29524
57. Benjamin S, Prakasan P, Sreedharan S, Wright ADG, Spener F (2015) Pros and cons of CLA consumption: an insight from clinical evidences. Nutr Metab (Lond) 12(1):4. https://doi.org/10.1186/1743-7075-12-4
58. Li K, Sinclair AJ, Zhao F, Li D (2018) Uncommon fatty acids and cardiometabolic health. Nutrients 10(10):1559. https://doi.org/10.3390/nu10101559
59. Wannamethee SG, Jefferis BJ, Lennon L, Papacosta O, Whincup PH, Hingorani AD (2018) Serum conjugated linoleic acid and risk of incident heart failure in older men: the british regional heart study. J Am Heart Assoc 7(1). https://doi.org/10.1161/JAHA.117.006653
60. Zárate R, el Jaber-Vazdekis N, Tejera N, Pérez JA, Rodríguez C (2017) Significance of long chain polyunsaturated fatty acids in human health. Clin Transl Med 6(1). https://doi.org/10.1186/s40169-017-0153-6
61. Zhuang P, Wang W, Wang J, Zhang Y, Jiao J (2019) Polyunsaturated fatty acids intake, omega-6/omega-3 ratio and mortality: findings from two independent nationwide cohorts. Clin Nutr 38(2):848–855. https://doi.org/10.1016/j.clnu.2018.02.019
62. Ilyasoglu H (2017) Production of structured lipid with a low omega-6/omega-3 fatty acids ratio by enzymatic interesterification. Grasas y Aceites 68(2):191. https://doi.org/10.3989/gya.0565161
63. Feltes MMC, De Oliveira Pitol L, Gomes Correia JF, Grimaldi R, Block JM, Ninow JL (2009) Incorporation of medium chain fatty acids into fish oil triglycerides by chemical and enzymatic interesterification. Grasas y Aceites 60(2):168–176. https://doi.org/10.3989/gya.074708

Epilogue

Concluding Remarks

The replacement of industrial trans-fats has been primarily done with interesterified fats rich in palmitic acid, using palm oil and its fractions combined with various vegetable oil sources such as soybean, cottonseed, corn, sunflower, among others [1, 2]. However, the use of palm oil has been criticized due to growing concerns about potential environmental and social implications associated with the expansion of cultivation areas. Regarding technological aspects, it is worth noting that during its fractionation and refining process, undesirable by-products such as DAGs and contaminants like 3-monochloropropane-1,2-diol (3-MCPD) esters and glycidyl esters can be generated, these substances are formed from the reaction of TAGs and DAGs with chlorides and may be associated with toxicological effects (carcinogenic and/or genotoxic), this raises serious concerns about their presence in foods, especially in infant formulas, which are often the sole nutritional source for infants [3, 4]. According to a recent pilot study on the presence of 3-MCPD esters and glycidyl ethers in commercial infant formulas, there have been reductions in chloropropanol esters compared to previous investigations available in the literature; however, there are still no strategies for the complete elimination of these contaminants [3, 5–7]. Therefore, other potential applications should also be explored, such as harnessing new sources of low-cost hardfats and vegetable oils, rich in essential FAs and/or of nutritional interest (bioactive compounds), in the synthesis of SLs. This would contribute positively to both technological and nutritional applications, remaining economically viable and having the potential for use in nutraceuticals and functional foods [8–10].

However, it is worth noting that regarding the application of SLs in foods, there is no labeling requirement for products containing this structuring fat. This is because it is challenging to accurately estimate the total dietary intake of SLs by a population. Therefore, assessing the SL content in foods has become a current challenge [2, 10, 11]. It is also emphasized that the operational cost of enzymatic interesterification is

V. Alves et al., *Chemical and Enzymatic Interesterification for Food Lipid Production*, Chemistry of Foods, https://doi.org/10.1007/978-3-031-67405-1

high, requiring immobilized enzymes due to the low reaction stability. This difficulty hinders its large-scale implementation by industries, making it necessary to implement sustainable and low-cost enzymatic processes. This could be achieved through the utilization of oils and fats extracted from by-products of agro-industry as raw materials [10].

Another factor to consider is that although extensive studies have been conducted on the physicochemical and functional properties of SLs with promising results regarding their applicability in foods, studies on the lipid metabolism of SLs after consumption are limited. Therefore, it is necessary to understand how the restructuring of TAGs can interfere/act in metabolism and what their long-term health effects are. This is crucial since postprandial effects of SLs are investigated only in acute in vivo animal studies, limiting their applications in humans [11, 12].

Conflict of Interest
The authors declare no conflict of interest.

References

1. Lefevre M, Mensink RP, Kris-Etherton PM, Petersen B, Smith K, Flickinger BD (2012) Predicted changes in fatty acid intakes, plasma lipids, and cardiovascular disease risk following replacement of *trans* fatty acid-containing soybean oil with application-appropriate alternatives. Lipids 47(10):951–962. https://doi.org/10.1007/s11745-012-3705-y
2. Mills CE, Hall WL, Berry SEE (2017) What are interesterified fats and should we be worried about them in our diet? Nutr Bull 42(2):153–158. https://doi.org/10.1111/nbu.12264
3. Beekman JC, Granvogl M, MacMahon S (2019) Analysis and occurrence of mcpd and glycidyl esters in infant formulas and other complex food matrices, pp 67–90. https://doi.org/10.1021/bk-2019-1306.ch005
4. Sim BI et al (2020) Mitigation of 3-MCPD esters and glycidyl esters during the physical refining process of palm oil by micro and macro laboratory scale refining. Food Chem 328:127147. https://doi.org/10.1016/j.foodchem.2020.127147
5. Becalski A, Zhao T, Feng S, Lau BP-Y (2015) A pilot survey of 2- and 3-monochloropropanediol and glycidol fatty acid esters in baby formula on the canadian market 2012–2013. J Food Compos Anal 44:111–114. https://doi.org/10.1016/j.jfca.2015.08.004
6. Wöhrlin F, Fry H, Lahrssen-Wiederholt M, Preiß-Weigert A (2015) Occurrence of fatty acid esters of 3-MCPD, 2-MCPD and glycidol in infant formula. Food Additiv Contamin: Part A 32(11):1810–1822. https://doi.org/10.1080/19440049.2015.1071497
7. Nguyen KH, Fromberg A (2020) Monochloropropanediol and glycidyl esters in infant formula and baby food products on the danish market: occurrence and preliminary risk assessment. Food Control 110:106980. https://doi.org/10.1016/j.foodcont.2019.106980

8. He Y, Li J, Guo Z, Chen B (2018) Synthesis of novel medium-long-medium type structured lipids from microalgae oil via two-step enzymatic reactions. Process Biochem 68:108–116. https://doi.org/10.1016/j.procbio.2018.02.005

9. Mota DA et al (2020) Production of low-calorie structured lipids from spent coffee grounds or olive pomace crude oils catalyzed by immobilized lipase in magnetic nanoparticles. Bioresour Technol 307:123223. https://doi.org/10.1016/j.biortech.2020.123223

10. Sivakanthan S, Madhujith T (2020) Current trends in applications of enzymatic interesterification of fats and oils: a review. LWT 132:109880. https://doi.org/10.1016/j.lwt.2020.109880

11. Mensink RP, Sanders TA, Baer DJ, Hayes K, Howles PN, Marangoni A (2016) The increasing use of interesterified lipids in the food supply and their effects on health parameters. Adv Nutr 7(4):719–729. https://doi.org/10.3945/an.115.009662

12. Shagholian M, Goli SAH, Shirvani A, Agha-Ghazvini MR, Asgary S (2019) Liver and serum lipids in Wistar rats fed a novel structured lipid containing conjugated linoleic acid and conjugated linolenic acid. Grasas Aceites 70(2):307. https://doi.org/10.3989/gya.0582181